First published in 2015 by Voyageur Press,
an imprint of Quarto Publishing Group USA Inc.,
400 First Avenue North, Suite 400, Minneapolis, MN
55401 USA

The information in this book is true and complete
to the best of our knowledge. All recommendations
are made without any guarantee on the part of the
author or Publisher, who also disclaims any liability
incurred in connection with the use of this data or
specific details.

We recognize, further, that some words, model
names, and designations mentioned herein are the
property of the trademark holder. We use them for
identification purposes only. This is not an official
publication.

Voyageur Press titles are also available at discounts
in bulk quantity for industrial or sales-promotional
use. For details write to Special Sales Manager at
Quarto Publishing Group USA Inc., 400 First Avenue
North, Suite 400, Minneapolis, MN 55401 USA.

To find out more about our books,
visit us online at www.voyageurpress.com.

ISBN: 978-0-7603-4563-4

Library of Congress Cataloging-in-Publication Data

Tweten, Wendy, author.
    Gardening for the homebrewer : plants for making
    beer, wine, gruit, cider, perry, and more / by
    Wendy Tweten and Debbie Teashon.
        pages cm
    Includes index.
    ISBN 978-0-7603-4563-4 (sc)
    1. Food crops. 2. Plants, Edible. 3. Plants, Useful.
    4. Beer. 5. Wine and wine making. I. Teashon,
    Debbie, author. II. Title. III. Title: Plants for mak-
    ing beer, wine, gruit, cider, perry, and more.
    SB175.T84 2015
    635'.2—dc23
    2015007057

Acquisitions Editor: Thom O'Hearn
Project Manager: Jordan Wiklund
Art Director: Cindy Samargia Laun
Book Design: Brad Norr
Book Layout: Wendy Holdman

Printed in China

10 9 8 7 6 5 4 3 2 1

GARDENING *for the*
HOMEBREWER

GROW AND PROCESS PLANTS FOR MAKING BEER, WINE, GRUIT, CIDER, PERRY, AND MORE

WENDY TWETEN AND DEBBIE TEASHON

Voyageur
Press

Contents

Introduction

For those of you who weren't there, TV ads of the 1960s were heavy on cartoon characters pitching breakfast cereals, tank-solid cars caressed by girls with pageboy hairdos, and beer with heads of foam that exploded out of the glass and flowed down the sides like golden volcanoes.

Decades later, there are still similar commercials. However, many of us have come to understand that beer, wine, cider, and liqueurs are more than a buzz: They're art and science rolled into intriguing and nuanced libations. Just as with fine food, those that enjoy a craft beer or well-made wine know that the ingredients make a difference. So it makes sense that if you brew your own, at some point you may wonder: Can I also grow my own?

The good news in a nutshell is a resounding "yes!" Beginners as well as experts can be proud of their forays into gardening just as they were in their early days of brewing. In other words, even a first effort is likely to produce an ample harvest, just as your first few batches provided a satisfying beverage. Of course, you can also end up with dead plants or a drain pour, but only if you're careless!

Therein lies the beauty of either hobby; these are pursuits of personal enrichment that have only a few simple rules for success. Here's the secret of each: When fermenting, sanitize everything to avoid brew-ruining contamination, and, in gardening, site plants correctly and don't let them dry out before they're established. If you can remember these guidelines, every step is a new adventure!

Brewing and gardening are good companions. After all, both are natural forces set into motion, then guided by the human hand. In the garden, fruit ferments naturally. Since fermentation relies on the garden, why not take control of the flavors that go into your brews by growing them at home? Well-crafted adult beverages have depth and refinement. One of the most important elements—freshness—is as elusive as it is esteemed. Elusive, that is, unless you're harvesting your ingredients from your own backyard.

We hope this book will plant a seed. A brewer's garden can be so much more than a production plot. With a little forethought, it can be a source of relaxation and pride. When the weather is fine, it's an outdoor room. When the year turns cold, your garden becomes ever-changing art outside the windows of your warm house. As two hobbies grow into one, your efforts in the garden and with the fermenter will become second nature. May you gain coordination as your confidence grows, while losing none of your enthusiasm!

CHAPTER 1

Before You Begin
Tools, Materials, and More

R emember playing outside when you were a kid? Climbing trees? Building forts? Turning over rocks to see what crept beneath? It was fun, wasn't it? Well, then you went and grew up. Try having all that alfresco fun as an adult and you're going to get a lot of funny looks. But there's one way to play in your back yard without becoming the neighborhood eccentric: get out and garden.

Not only is gardening enjoyable on its own, it's a natural accompaniment to fermenting. Grow it fresh, brew it fresh, and reap the rewards of both hobbies in a most satisfying way. Name two other mutual pursuits whose industry results in an invitation to relax. Do a little harvest, make a little must, and sit down tonight (or in a few months, unless there's homebrew already in your fridge).

PICKING YOUR PLANTS

Before you get out there planting and tending your garden, there's the planning. The most basic consideration is selecting plants that will work in your climate. There's a reason people don't grow avocados in Seattle or apples in Florida. Sure, you can experiment with out-of-region plants, but really, with all the great things you can grow, why bother? Seek success

with climate-appropriate plants and grow them well. A little investigation up front is worth months (or years) of frustration.

USDA Hardiness Zones are a solid starting point that can narrow your initial plant list before you move on to other considerations. The zones, developed by the United States Department of Agriculture, identify the average minimum temperature likely in a particular area. Be aware, though, that in this case hardiness means only tolerance of a certain range of low temperatures. It doesn't mean the plant is necessarily a rugged performer in that region. For instance, apricot trees are hardy in USDA Zone 8, but in coastal Northwest Zone 8, the buds of most apricots are killed by spring frosts and the trees fall victim to various diseases due to dampness.

The Zones also don't reveal a plant's tolerance or need for summer heat, humidity,

Above and opposite: Whether you draw inspiration for your garden from the garden center or the great outdoors, the options are endless to craft a space to suit your home and brewing needs.

insulating snow cover, number of chill hours, or day length—all big factors in plant performance. For these reasons and more, privately-owned local nurseries, County Extension Services, and fellow gardeners are excellent sources of advice before you plant. Many Extension offices of state land grant colleges also have websites with expert gardening information. Community garden tours that let you get up close and personal with plants and varietals that perform where you live are another option.

Now you know some plants that can grow in your garden. But wait just a minute. Filling a garden is about more than what you can grow; it's also a matter of what you shouldn't grow. One of the most com-

mon mistakes made by gardeners—green thumbs as well as greenhorns—is not taking into account a plant's true height and spread at maturity. The size listed on the nursery tag of long-lived plants is an estimation of the size at ten years, not necessarily when full-grown. This makes a big difference not just when spacing plants from each other but from a structure or a driveway.

That's not the only pitfall. Some plants are just too willing to perform. These are the dreaded invasives. Do you like mint? If you planted it in your garden, let's hope so, for you will never be rid of it. How about some horehound to brew homegrown gruit beer? Allow it to reseed just once and gird your loins for an onslaught of seedlings. That doesn't mean you have to exile these lovely bullies from your garden altogether, just take extra care by confining them to pots or removing seedheads before they ripen.

You and your plants: It's a symbiotic relationship. The plants must serve your needs, but you, in return, must serve theirs. This means siting your little green friends based on what they require. Become familiar with your property's microclimates; the little places that hold heat, invite frost, or are the last place to lose the sun as the days shorten. Put these mini-zones to work for you, placing heat-loving plants in a sunny, south-facing situation and those that appreciate a bit of moisture in that consistently damp area. When choosing your plants, do your research and let the plant tag be your guide.

Decoding the Plant Profiles

In this book, you'll find plant profiles for just about everything we discuss. They'll provide the nuts and bolts needed for you to figure out whether the plant will be a good fit for you or not. They'll also provide important growing information and other things to keep in mind once the plant is in the ground.

PLANT NAME: These are the plants you know and love by their common name: i.e., cucumber, tomato, apple, lavender.

BOTANICAL NAME: Common names are well and good most of the time, but sometimes it's nice to know exactly what you're talking about. Botanical nomenclature provides a scientific identity for every plant known to mankind. It provides a standard by which plants are recognized throughout the world, which serves two important purposes for the home gardener. First, it heads off any confusion. If a friend offers you a piece of her favorite daisy plant, it's helpful to know which of at least a dozen different genera known as "daisy" you'll be receiving. Second, it provides insight into plants within a genera. Take lavender, for example: The name *Lavendula augustifolia* (English lavender) tells you your species is hardier and much more fragrant than *Lavendula stoechas* (Spanish lavender).

PLANT TYPE: An annual grows from seed, sets seed, and dies within a single growing season. A biennial grows foliage the first year; the second year it flowers, sets seed, and dies. A perennial lives for several years, some for only a few, some for a century or more. Most perennials die to the ground in the dormant season and come back from the roots in spring.

A shrub (or bush) is a woody plant up to 10 feet tall with several stems rising from the ground. A tree is woody plant, generally 10 feet or taller (except dwarf varieties), with a single trunk and lateral branches. An evergreen doesn't lose its foliage in the dormant season as a deciduous plant does. Most evergreens are coniferous (coned), needled plants, but some are broadleaf, having leaves rather than needles. On the other hand, not every coniferous, needled plant is evergreen; conifers such as larch and dawn redwood lose their needles in autumn. And not every plant with cones has needles, notably the gingko.

USDA ZONES: Every region of the United States has a USDA-assigned Hardiness Zone, which reflects the area's average minimum temperatures. Plants are rated according to the USDA Zones in which they will survive.

HEIGHT: It's important to keep in mind the size listed on the nursery tag of long-lived plants is an estimation of the size at ten years, not necessarily when full-grown.

SOIL: Loam is a good garden soil of clay and sand containing humus. Humus is decomposed organic matter in the soil. A boggy soil is one that never completely dries out. Peaty soil contains large quantities of peat, which holds moisture and tips the pH into the acid range. Well-drained soil does not hold moisture for long.

LIGHT: Full sun typically means at least six hours per day of direct sunlight. However, many plants, such as vegetables, do better with at least ten hours.

WATER: Some plants, such as many herbs, prefer soil that dries out completely between waterings. These low-water plants can even withstand extended periods of drought. High-water plants need soil that is consistently moist. Plants that prefer boggy ground must have soil that is consistently wet. Between the extremes of wet and dry lie the majority of plants that are most comfortable with regular water, an inch or so per week, and soil that becomes dry only briefly, if at all.

PROPAGATE BY: Plants propagated by seed can be either direct-seeded—planted directly into the ground—or grown in pots

and planted out as starts. Any method of propagation other than by seed is asexual propagation, which produces clones of the original plant. Division creates new plants by dividing the rootball, either by slicing out a section with a shovel, or by digging up the entire clump and cutting it into pieces with a shovel, pruning saw, axe, knife, or loppers.

Cuttings are most often taken of stems. Cuttings are typically dipped in rooting hormone, placed in a soilless mix such as half peat moss and half perlite. The mix is kept consistently moist. The time of year and size of the cuttings is determined by the plant to be propagated.

Layering is a long-term but easy process for certain plants. A branch still attached is nicked below a node (the point from where the leaves grow) and that point is bent down and buried just below the soil surface where it's held in place by, usually, a rock. If kept moist, the node should root. Once the area is well rooted—which can take up to a year—the new plant is cut free, but not dug out. In several months to a year, the layered plant should be well established and ready to dig and replant.

Many fruit trees are propagated by grafting or budding. Grafting brings together a branch from the desired tree (called a scion) onto the roots of a tree of the (usually) same species. Budding attaches a bud from the donor tree to a larger rootstock. Scions and buds can also be attached to established trees.

SOIL

Now that you've narrowed down what to plant, let's see what you've got to plant it in. Is your ground soil, or is it just plain dirt? If it's easy to dig and neither too soggy nor too sandy, chances are you have something to work with. However, a load of well-rotted compost or manure tilled into the growing area—not just the planting hole—is almost always a worthwhile investment. If your soil is more like rocky hardpan, consider rising above it with bermed beds of quality topsoil mounded on top, or giving up on it altogether and growing in containers. With any garden, there is preparation involved. If your property isn't blessed with rich, dark earth, you can enhance it by adding soil amendments every year. Garden compost, chicken or livestock manure, or mushroom compost all add humus to the dirt. If you make a habit of improving your soil every year with nutritious mulch, your soil can become the envy of the neighborhood in a few years' time.

Note: Many herbs actually prefer lean, dry conditions, and produce better flavors in infertile soil. These low-water, hot-summer herbs include thyme, rosemary, oregano, sage, anise, caraway, and bay laurel. It doesn't matter if they're growing in Idaho or Alabama; they long for the sun, sandy soil, and warm winters of the Mediterranean. If you wish to grow them well, make their dreams come true with at least six to eight hours of full sun on the south side of the house in northern climates. Although their roots require excellent drainage, a soil amended with compost actually helps create a loose, light, loam. Newly planted herbs need to be watered until they're established, as evidenced by strong growth. At this point they demand water only during dry spells and heat waves. A little complete fertilizer or compost worked into the soil around the roots at the start of spring is all these herbs need to stay healthy. With their craving for warmth and drought, it's no surprise Mediterranean herbs are naturals for containers.

If you'll be building a new bed out of a weed-ravaged piece of ground, there are several ways to turn it into a lovely, loamy, virgin garden plot. The quickest way is to spray a foliar-acting herbicide and proceed once the weeds have died, which can be in one to three months, depending on the herbicide used, time of year, and type of weeds. For those who wish to keep chemicals out of the equation, the same effect can be achieved much more slowly with a four-inch or more layer of mulch or cardboard or four-sheets-thick or more layer of newspaper, completely covering every inch of the feral ground and topped by a 3-inch or more layer of leaves, compost, or manure. Up to a year might be required until all the weeds are truly dead. Some, such as horsetail, may simply refuse to die . . . ever. Tilling without first killing existing plant life will only propagate many of the weeds you're trying to eliminate. If

Make-It-Yourself Compost

When making your own compost, allow it to completely break down in its own pile before incorporating in the soil. Plant matter allowed to rot in soil with seedlings or other young plants will tie up nitrogen needed by the young crop. However, pre-composted material can be used as a mulch or a fall amendment in a bed that will lie fallow until spring. When mixing fresh material into the ground, a sprinkle of an organic nitrogen source, such as blood meal, poultry manure, or alfalfa meal, will help the compost break down faster and prevent the heartbreak of nitrogen deficiency.

The Difference between a Soil Amendment and Mulch

It's important not to confuse the terms "soil amendment" and "mulch." Simply stated, the difference between the two is about where it is, not what it is. A soil amendment is anything worked into the soil to improve it. Amendments include compost, alfalfa pellets, coir fiber, and lime. Mulch stays on the surface of the soil. Mulches include organic and inorganic materials such as wood chips, gravel, straw, bark, and leaves; these needn't be broken down before being shoveled on. Well-rotted organics, notably compost and manure, can be used as either a soil amendment or a mulch. Most materials should be allowed to break down before they are used as soil amendments, especially tree-based matter such as wood chips, sawdust, and leaves, which, as it rots, will tie up nitrogen needed by the plants. If the beds are to be planted within a few days or weeks, use only manure that has sat for at least six months to allow excess ammonia and salts to leach away, and to decrease pathogens. Fresh manure can be put directly on or in beds if they won't be planted for at least three or four months.

turning a lawn into a garden, strip off the sod in sheets and compost it. At any stage, a 3- to 4-inch mulch on top of the soil will discourage weed seeds from germinating.

Once the ground is vacant, till it or simply start planting. If you have bare ground and several months before you intend to plant, consider sowing a cover crop (also known as green manure) in spring or fall. Cover crops include, but are not limited to, crimson clover, fava, Austrian field peas, buckwheat, alfalfa, or rye and vetch. When the plants begin to flower, plough them under to improve the soil. Once the cover crop has been turned under, wait three weeks before planting to allow the material to rot and yourself time to apprehend any clumps of the crop that struggle to re-root.

Fall is a good time to amend your garden with cover crop, compost, manure, and other soil conditioners. By spring, the ground should be settled, enriched, and ready to grow—a highly desirable condition known as having good "tilth." Once garden soil has reached this enviable state, annual tilling is not only unnecessary, it can be harmful to the structure and microbial balance of the soil. Of course, if you're growing a cover crop, tilling is required, unless you take an annual crop such as fava, mow it low and let the roots die; however, you then lose the advantage of all that green material returned to the soil. In lieu of cover crop, add a 3- to 4-inch blanket of compost or manure to the top of the beds each year.

TOOLS AND EQUIPMENT

As a budding gardener, you have lots of exciting options and opportunities ahead. But one thing is as certain as it is mundane: you're going to need a shovel. If your garden will consist of a window box, all you really need is an old spoon and a watering can, but if you're planning to get serious about gardening you'll need some basic equipment.

Tools and Equipment

Shovel: You will need at least one standard round-point shovel, although two or three are better.

Trowel: for planting and digging shallow holes or those in close quarters. Narrow-bladed trowels can be handy for prying out the roots of tap-rooted weeds and small stones from rocky soil.

Soil (bowhead) rake: for spreading mulch, leveling soil, clearing the soil surface, or rebuilding mounded raised beds after tilling. A metal, bamboo, or plastic leaf/lawn rake is also useful for cleaning soil and collecting leaves.

Oscillating hoe: This cuts weeds off at the crown or uproots them entirely when slid back and forth just under the surface of the soil. Easy to work around plants, it won't remove tap-rooted plants like dandelion; to remove deeply rooted weeds such as these, try a slender, fork-tongued asparagus knife (also known as a fishtail weeder).

Hand pruner: for pruning, deadheading, taking cuttings, and harvesting.

Lopper: for pruning with leverage.

Pruning saw: If you have trees or large shrubs, a pruning saw is handier than a carpenter's saw.

Hoses: Invest in good quality, non-kinking hoses. Having hoses in various areas of the garden is much more convenient than dragging one hose everywhere. A screw-on hose nozzle with a simple on/off switch saves water or trips back to the faucet if you need to pause while watering.

Sprinkler: better yet, soaker hoses put the water right where you want it.

Wheelbarrow: It's hard to beat an old-fashioned wheelbarrow for moving stuff in, around, and out of the garden. If your garden's small, a bucket or large, plastic nursery pot makes a good weed collector in lieu of a wheelbarrow. A handtruck can be a back-saver if there are large pots or bags of potting soil to transport.

Gardening gloves: Don't be a hero. Latex-dipped garden gloves are available at practically every nursery and home improvement store. They're inexpensive and can save your green thumbs from spines, stings, infections, and dirty fingernails.

Garden clogs: A quality pair of waterproof garden clogs will keep your feet dry and the dirt out of your living room, if you get in the habit of removing them at the door. Good garden clogs are comfortable, slip on and off without your bending over, and last for many years.

PLANTING

It's time to break in that new shovel! Trees and perennials grown in containers (those with a rootball) can be planted any time except during the hottest or coldest weather. During the dormant season, newly planted perennials may languish and rot in the cold, wet soil, so it's best to wait until the end of the dormant season when growth is starting. Bareroot trees and perennials are available in late winter or early spring and should be planted immediately. Seeds are variable as to the time of year and method of planting they prefer. Refer to the packet directions.

When planting a tree, dig the hole as deep and twice as wide as the rootball or bare roots to be planted. For perennials

To Stake or Not to Stake

Stake trees or other plants only if you have doubts about their ability to stand on their own. When allowed to move with the wind, a plant develops a sturdier trunk and root system. If your newly planted tree has a trunk that is too weak to stay upright, or if the roots are incapable of supporting a top-heavy plant, staking will be necessary. Pound two stakes into the ground, one to either side of the rootball. Do this while you can see the roots, if possible, to avoid damaging them. Tie the trunk with broad, soft ties such as nylon stockings, strips of fabric, or commercial ties from a garden center. Place the ties—one to either side—at the point where they hold the trunk in position. Leave enough slack for the trunk to move 2–3 inches in every direction. Remove the supports by the time the plant is a year old.

and annuals, dig the hole the size of the rootball plus an extra inch or two. The roots of bareroot plants should be soaked in a bucket of water for four hours (but no longer than a day) prior to planting. For potted plants, gently loosen the roots and settle the plant in. For all plants, refill the hole with the same soil that came out of it; amending only the soil that goes into the hole will encourage the new roots to grow around in that small space, enjoying the rich banquet you have provided and eschewing the comparatively dull dirt at the borders. If you're feeling frisky, the best course of all is to amend the entire bed.

Backfill the hole halfway with soil, pressing it in firmly. The top of the rootball should be level with the top of the ground. Pour water to the top of the remaining hole. When the water has soaked in, finish backfilling. Build a circular levee of soil around the dripline to trap future waterings over the roots.

SEEDING

The time to plan your seed-grown garden isn't spring—it's winter. Right after the new year, start checking seed catalogs and online sites and get your order in as soon as possible. In some cases, seeds can go in the ground as early as February. While it's all right to use seed that's been stored for a year or two in an airtight container in the crisper drawer of the refrigerator, fresh seed, dated that year, is always your best bet.

Research your seeds to determine when they should be planted and if they prefer being sown directly in the ground or started in pots indoors. A general rule of thumb is that seeds are planted no deeper than their diameter. This means that very fine seed should be planted at or barely below the surface of the soil. Larger seeds can be pushed into the ground ¼ to ½ inch deep, depending on their type and size.

If you plant your seeds in the correct season in a bed that gets plenty of sun and warmth, your only saboteurs are likely to be dry soil and pests. Slugs, rabbits, earwigs, cutworms, mice, and voles can all take a toll on seedlings. If you notice sprouts and seed-leaves neatly clipped off, or disappearing altogether, you may want to visit your garden with a flashlight in the dead of night to catch the perpetrator in the act. Although seedlings are tougher than they seem, one thing they cannot tolerate is dehydration. Don't let the seedbed dry out completely. On the other hand, consistently soggy soil can be just as deadly.

Seeding Indoors

Many annuals, vegetables, and perennials can be sown indoors to give the plants a head start. In general, the seeds are planted four to eight weeks before you wish to put them in the ground. Use small pots or cell packs. Sterile, soilless seed-starting mix is available at most nurseries and worth seeking out. Fill the pots with the mix and dampen it thoroughly. Hot water is taken up better than cold by the peat moss or coir fiber that serves as the base of most of these sterile mixes.

Plant the seeds according to packet directions. Put more seeds in each pot than you need; you'll thin to the best plant later. To avoid dislodging the seeds when watering, place the pots in a tray and water from the bottom. Remove the pots when the soil is moist. Many seeds don't need light to germinate, but they do need warmth. A seed-starting heat mat under the pots can be a great help in maintaining the correct temperature. Keep the mix moist; once the seeds begin the process of germination, even a brief spell of complete dryness is a death sentence.

On the day the first seedling breaks ground, bring on the light—and lots of it. Shop lights with fluorescent bulbs placed an inch or two above the little plants are a steady, relatively cool source of light. As the plants grow, raise the lights, always keeping them barely above the top leaf. Unless you are extremely disciplined, put the lights on a timer that keeps them on eighteen hours a day—say on at 5 a.m. and off at 11 p.m. To bring extra light to the plants, and make the mornings and evenings a bit darker for you, run sheets of aluminum foil from the edge of the lights to the base of the pots; these sheets are easily pulled aside for daily plant checks. The seedlings can be removed from the heat mat once most of them have germinated. Be cautious about using the heating mat and shop lights together, as the combination can create temperatures that will quickly dehydrate the growing medium.

As soon as possible, thin to the strongest seedlings, usually one per pot or cell. Once the true leaves show (the first leaves to appear are the seed leaves), begin feeding with a heavily diluted complete fertilizer every week or two. If the plants outgrow their pots before they can safely be planted outside, pot them up to a larger container using regular potting soil and a little time-release

fertilizer. Ten days to two weeks before they will be planted outside, begin to "harden off" the plants by setting them outside for a few hours each day. Begin with two or three hours in the shade on a relatively warm day, and gradually work up to a full day in the sun, or wherever it is they'll be planted. Once they're acclimated, it's time to put them in the ground.

Care after Planting

Don't let newly planted perennials, shrubs, or trees dry out—at least not during their first two or three summers in the ground. Soaker hoses are a convenient, inex-pensive, and low-tech way to keep water right where it's needed. For most plants, an annual mulch of compost or manure can be the difference between surviving and thriving. When mulching, keep the material from touching the trunk or stem. Feeding will also boost the performance of your plants, especially ones that are just becoming established. Most plants appreciate a complete fertilizer such as fish emulsion applied in spring. Plants in containers and those grown from seed may need more frequent feeding.

Regular watering during the growing season makes for healthy mature plants

Orchard trees with alpine strawberries

as well. If fruit is not your goal, cut off spent flowers (deadhead) after blooming to keep plants neat. Thin tree fruit as suggested under plant listings to improve the quality of the remaining fruit and prevent branches from breaking. When it comes to pruning, different plants require different styles, but dead wood can be removed any time the pruners are sharp, as the saying goes. If you have an old, overgrown tree, consider hiring an arborist certified by the International Society of Arboriculture to bully it into shape—probably over the course of three or more years.

BOUNTY WITH BEAUTY: PLANNING A PRODUCTIVE GARDEN

Nowhere is it written that you have to choose between a beautiful landscape and a productive garden. Don't hide your garden away! Plant it proudly in the front yard, or, better yet, make it the heart of an outdoor living area where you go to relax, read, entertain, and enjoy the company of your family. Just about any plant you can brew or ferment will do double-duty as an ornamental part of the landscape. A grape trellis is a natural living screen. A small fruit tree takes center stage with a blush of spring blossoms and bounty of summer fruit. Dense green globes of dwarf Sitka spruce slowly grow into pyramidal evergreen shrubs that are the basis of a distinctive seasonal specialty beer. An arbor shades a porch swing beneath a roof of hop

vines. Formal herb gardens, or a meadow of herbs, can include many of the ingredients for gruit beer and ancient ales.

Hops may be grown on an arbor overhanging a patio, provided you keep them far enough from seating areas to prevent them from enveloping your guests' heads in hops. A simple structure of columns holding up overhead beams that attach at their far ends to the house may not be a traditional hops support, but will create a fine grotto for summer lounging while still providing a harvest, especially if your hops receive plenty of light. Hops also can be used as a backdrop if grown vertically up posts or on a pole-and-wire system.

When it comes to cider, a 20-foot square of backyard can hold one standard apple tree, two semi-dwarfs, or four fully dwarf trees. Three dwarf apple trees leave room for another type of fruit tree to round out your cider-making possibilities. Depending on your zone, consider a dwarf peach, quince, or sour cherry—great cider additions all. Your three dwarf apple trees should, in two or three years, present you with enough fruit for about 9 gallons of fresh cider. If your climate is mild, give the king of cider apples, Kingston Black, a try. For your second tree, consider a bittersweet such as a vigorous Major or Medaille D'Or. Your third tree can round out your cider triumvirate with a sweet such as a Grimes Golden, which is aromatic and doubles as a culinary apple, or a Roxbury Russet with the same characteristics but also offers good storage and a tolerance for colder climates. In place of a sweet cider apple, you can plant your favorite culinary apple. Just be sure it ripens around the same time as its orchard-mates; this not only makes for convenient pressing but also helps ensure pollination since apples that ripen together generally bloom together.

Even a small garden space can produce ample amounts of brewing material. Espaliered fruit trees turn an abandoned sunny side yard into a fruit-producing area, for the width of only a few feet. If the area is wide enough, add some blueberry bushes, strawberry groundcovers, herbs, and a few strictly ornamental plants to turn it into a five-star side garden. To make it seem even less of an alleyway, run a gently curving path through. In a larger space, well-placed pots of anything from grasses to vegetables break up the space, and with a colorful bistro table and chairs, your brewer's garden doubles as a place to decompress after a long day of work. Add a bubbling fountain, aromatic lilies or honeysuckle, a glass of home-crafted, ice-cold, dark-as-night porter, and your senses are filled.

As a home grower, you have one big advantage over commercial brewers and fermenters: You can grow specific and specialized plant varieties that make the difference between a mediocre beverage and a memorable one. For instance, true cider and perry trees are rare in the United States, yet their fruit is the best way to ensure a quality product. Most stateside ciders and perries come from apples and pears that were bred not for fermenting but for eating out-of-hand. Happily, most of us can grow a cider tree in our home garden; one productive semi-dwarf tree will provide you with plenty of pome-fruit potables. Hops are another example: those used in commercial beers are generally chosen for their heartiness and productivity; less robust but distinctively flavored hops may be ignored. However, for a backyard grower, these specialty vines can often be grown at home with only a little more care than their professionally preferred kin.

Berries are equally well suited to the home garden. As we all know, berries don't travel well and their freshness is easily lost. In the landscape they are forgiving plants that deliver their taste of summer from backyard to freezer to fermenter. Rows of bramble berries can be worked into the landscape with aesthetically pleasing mown-lawn paths between. New varieties of raspberry can be grown in pots. Blueberry bushes are handsome, all-season plants that often throw fall color into the mix. Elderberry bushes make dense summer screens. Cranberries are true groundcover plants that, as with all creepers, require maintenance to discourage competition from weeds, but in the end give you a crop to show for your efforts.

Technically, an herb is any plant used for seasoning, medication, or fragrance. For our purposes, herbs will also be used to add layers of interest to alcohol. Herbs are

notoriously easy to grow. Give them well-drained soil, six to eight hours of sun a day, keep them trimmed (preferably by harvesting), and be prepared to replace them every few years when they decline. Most herbs don't need rich soil, but do appreciate a complete fertilizer each spring. Herbs are good candidates for container culture. As a bonus, herb-filled pots can be placed close to the house for convenient harvesting.

Meyer lemon bush

The Container (Balcony) Garden

Herbs are by no means the only plants that take to containers. With a little ingenuity, a collection of pots, and an apartment balcony with good sun, you can grow many of the plants mentioned in this book. True, the size of your harvest will be smaller than that of a backyard garden, so plan wisely to ensure you have the amount you need for whatever recipe you have in mind. Containers are also useful for experimenting with plants that may not be winter-hardy in your area: Meyer lemon trees can be container grown in almost any climate. In a portable pot they can be snowbirds, spending summers outdoors and winters indoors. For other plants, such as dwarf peaches, this mobility is equally handy. Peach trees can be moved under eaves or other shelter to keep peach leaf curl–abetting rain off the emerging leaves in late winter, after which the plants can go into a sunny spot where their ripening fruit will be enjoyed as a garden ornament, and then as a fermentable delicacy.

When choosing containers themselves, look for those with a least one drainage hole. Keep in mind the weight of the container plus the moist potting soil if you need to move it. Good quality, lightweight containers of fiberglass and resin are a popular alternative to handsome, though heavy, ceramics. Weighty containers are a good choice for trees and tall plants at risk of blowing over. If you favor hefty planters, invest in a handtruck to prevent a slipped disc. Finally, if your pot will sit on a wood deck or other surface that might be damaged by constant moisture, put three or four pot feet around the bottom edge of the container. This also makes the bottom of the pot less attractive to slugs, snails, and earwigs.

The potting soil you use should be a reputable commercial mix, with or without time-release fertilizer. Fill the pot to the point where the plants will sit at the correct height. A little alfalfa meal makes a good soil amendment, delivering organic nutrition to roots in this confined space. Now add the plants and finish filling with potting mix, leaving plenty of room for water. Pour in water until it runs from the drainage holes. Plants in pots need more water than those in the ground. At the height of summer they may need to be watered once a day—or even more—depending on the plant, the site, the size of the pot, and how thoroughly you watered last time. Containerized plants also need more frequent feedings: mix time-release pellets into the soil or water with a liquid fertilizer every couple of weeks.

Whether you're cultivating acreage or a few square feet, your brewing garden is a place to kick back and let your fancy turn lightly to thoughts of delicious things to drink. Crack open a bottle of liquid gold lovingly put away the previous fall. Let the birds sing. Let the bees drone from flower to flower, pollinating the coming harvest. What could be finer than relaxing in the midst of this intoxicating environment, enjoying the labors of your fruit?

CHAPTER 2

The Big Beer Ingredients

Growing Hops, Malt, and Other Grains

I s there anything friendlier than a glass of beer? Pull up a stool and tell me about your day. Let's get the gang together for a cold one after work. Time for a clambake? Don't forget the beer. Whether poured by a barmaid in a place where everybody knows your name, or—even better—served from your own homegrown, home-brewed stash, an amber-colored ale lolling beneath a frothy white head signals that happy hour is here. We're all friends together; bring on the beer!

Casual, yes, but beer is far from unsophisticated. Just because no one expects you to smell the cap and swish a sip around in your mouth when the bottle arrives doesn't mean brewed equals boring. When it comes to nuance, beer can hold its own against fine wines. Whether you favor the refreshing crispness of a European lager, the hoppy happiness of an IPA, or the dark voluptuousness of a stout, there's a beer for nearly every taste. But there's more to brewing than malt and hops; from pumpkin to lemon balm, the list of uncommon additions to the beverage is as odd as it is endless.

In fact, the great diversity of beer makes it attractive not only to those who enjoy a good brew, but to those who wish to create one. As a hobby, the practice is not hard to learn, and, with experience, can result in beer that is better than the average selection available in the coolers of the local grocery. Brewing is a frugal avocation too. After the initial outlay for the equipment, the cost of homebrewed can be a fraction of that of commercially produced beers. Best of all, no one has more friends than the man or woman who makes beer.

So why not take brewing one step further? For the brewer who is adventurous or set on making a truly local beer, there is no greater source of pride than using ingredients from your own backyard garden. From hops and barley to pumpkin and peppers, and herbs of all types, homebrewed is a natural segue to homegrown.

HOPS

Hobbyists have one big advantage over corporate breweries: Small batches lend themselves to fresh ingredients and encourage experimentation. After all, the dawn of beer saw brewers using everything from nettles to wormwood. Many of the plants used to make beer can be easily grown in a home garden. Perennial hops, for instance, are incorrigible extroverts. Their botanical name, *Humulus lupulus*, gives a clue to their untamed nature. The genus name *lupulus* means "small wolf," referring to the Romans' observation that the plants are quite capable of swallowing up their garden neighbors. Give hops plenty of sun, enough water, and the slightest excuse, and they will run wild across the fields and menace passers-by with stems covered in tiny, flesh-grabbing hooks.

For gardeners who admire ambition in their plants and know better than to rub against the bristly bines, hops are a beautiful, useful addition to the garden. The bines will sprawl if unsupported and are best grown vertically. When trained up trellises, poles, or stringed supports, they can stretch to 20 feet or more. In summer, the handsome three-lobed leaves are joined on female plants by papery, cone-like flowers properly termed *strobili*.

Within these flowers lies the essence of modern beer. Small glands between the scale-like bracts of the female hops flowers produce a yellow powder called *lupulin*. This resinous substance contains alpha and beta acids that give beer its distinctive flavor and acts as a preservative. Hops are dioecious—each plant either male or female. Only females produce cones. In hopyards, male strays are removed from the premises faster than leather-jacketed lads who call on the sheriff's daughter. The males are easily unmasked; upon blooming the male flowers are merely flowers, very different from the papery inflorescences of the female.

Fortunately for the backyard grower, female plants are the only sex commonly sold. Hops plants are also easily shared, and a piece of female hops rhizome donated by a fellow gardener will yield another female plant. Male plants are not necessary to female flowering; in fact, if the females are fertilized, the resulting seeds make the cones useless for beer production. The flowers of hops make beer, not the fruit.

Hops

BOTANICAL NAME: *Humulus lupulus*

PLANT TYPE: perennial vine

USDA ZONES: 3–8

HEIGHT: 20 feet or more with trellising

SOIL: well-drained loam

LIGHT: full sun

WATER: regular

GROWTH HABIT: vining bines. The plants should be trellised.

PROPAGATE BY: rhizomes. Plant the rhizomes vertically with the buds facing up, or horizontally 1 to 2 inches deep after the threat of frost is past.

SPACING: 3 feet between plants of the same cultivar or 5 feet between different varieties

YEARS TO BEARING: one to three. The first year's crop is usually weak. Generally, hops plants take three years to produce a brew-worthy harvest. Yields continue to increase over the next several years.

PRUNING: Failure to prune results in badly tangled bines and difficulty in harvesting. Train three to four main bines up strings on a hops trellis. As the season progresses, cut out any bines that scramble along the ground. At the end of the growing season, prune the vines to 2–3 feet.

HARVEST: From midsummer on, check the cones regularly for ripeness. Indicators are a pleasant aroma, papery feel, and—in some varieties—a change in color from green to yellow-green. Ripe cones feel dry and will leave behind a slightly sticky lupulin residue upon handling. Begin picking at the top of the plants where the cones are likely to be ready soonest. On a single plant, the cones mature at different rates, so be prepared to harvest every few days until the cropping is complete, usually by the end of September.

NOTES: A single well-grown hops plant can yield up to 6 pounds of fresh cones—plenty for most homebrewers. Fertilize first-year plants once a month with a balanced fertilizer such as 10-10-10. For established plants, apply the same fertilizer twice a month from emergence of the cones until flowering. A mulch of compost or manure applied once a year is helpful.

BEST USED IN: beer, cider

Hops by Climate

While hops are extremely accommodating garden plants in regions where they grow well, they do not grow equally well in all regions. Hops plants are cold hardy and can be grown from Canada to northern Mexico. However, survival and production are two different things. These plants appreciate as many hours of sunlight as they can get. Hops require a minimum of fifteen hours of daylight for optimum flowering, which puts them at their best at or above 35 degrees latitude. This range excludes much of the lower third of the United States, although intrepid hop growers report success in such lower Zone 8 states as Florida, North Carolina, Texas, and New Mexico. In hot regions, hops will need less direct sun and more water. Recommended hops varieties for hot regions include Cascade, Centennial, Chinook Clusters, Crystal, Galena, Horizon, Magnum, Newport, Nugget, and Summit.

In states with high humidity, including North Carolina, Florida, and much of New England and the Midwest, hops may suffer from powdery and downy mildew. Varieties often grown in these areas are Cascade, Centennial, Columbus, Chinook, Glacier, Golding, Hallertau, Magnum, Mt. Hood, Northern Brewer, Nugget, Perle, Saaz, Summit, Tettnang, and Williamette. Hops handle cold well, but they do require a minimum of 120 frost-free days to produce cones. Growers in the frigid Midwest have good results with Cascade, Centennial, Chinook, Columbus, Galena, Glacier, Mt. Hood, Northern Brewer, Nugget, Sterling, Willamette, and Zeus. The best advice when growing hops in the home garden is this: keep what performs well, produces wantonly, and tastes good; replace the rest.

If you live in the Pacific Northwest, rejoice! Washington, Oregon, and Idaho are the big three of hops-producing states. Home growers in these states and other areas with mild temperatures and low humidity can choose from such favorites as Cascade, Centennial, Newport, Sterling,

Ornamental Hops

If you have the space for yet another hops plant, consider the lovely golden hops, *Humulus lupulus* 'Aureus'. Its name refers to the color of its leaves, which are more of a bright acid green than gold. Like its green-leaved relatives, this vigorous climber produces decorative, papery cones in the summer. The cones are not usually used in brewing, but since golden hops is a genetic mutation of beer hops, their use is at least a possibility.

Talk Hops Like a Pro

Bine: a plant with climbing shoots that circle around an object to climb. In the case of some bines, including hops, backward-facing bristles help the stems cling. They are distinct from vines which ascend via suckers or tendrils. Hops (and many other bining plants) twine clockwise, as opposed to the counterclockwise direction of most vines.

Crown: where the upper structures of a plant meet the roots at or just above the ground. In the case of hops, which die to the ground each winter, the crown is where new growth emerges in the spring.

Dioecious: plant species which have male and female flowers on separate plants. Examples of dioecious plants are hops, holly, cannabis, kiwi, ginkgo, and yew.

Lupulin: a resinous, yellow powder produced by female hops flowers and used by humans to preserve and flavor beer.

Oast: a structure ranging in size from a barn to a box used to dry hops via circulated heat or air.

Rhizome: underground stems with the appearance of swollen roots. Hops rhizomes can be dug up and divided to propagate new plants.

Strobilus: female hops flower. Strobili are cone-like inflorescences.

and Nugget, or can experiment with any other hop variety that tickles their fancy. The choices are head-spinning: There are more than 100 varieties of hops from across the globe. Europe, Asia, the Czech Republic, New Zealand, and Australia have all added to the gene pool. The key ancestor for today's bittering cultivars came from Manitoba, Canada. And new selections keep coming. Hops hybridizers delight in introducing new varieties. If the botanists have their way, the next decade will see hops that are daylight-neutral, mildew resistant, early maturing, and even higher yielding.

Fortunately, among the dozens of varieties available to the homeowner, there are already varieties of hops that fare well against heat, mildew, pests, and diseases. At the top of most lists is aromatic, all-purpose Cascade, which has been performing reliably in a wide range of climates for sixty years.

Hops for Flavor

Let's not lose sight of the fact that choosing hops is about more than horticulture; it's about taste. Brewers know that different hops impart different characteristics. For example, hops can be used for bitterness, aroma, or both. Which hops are best suited for each purpose relates to the alpha and beta acids contained in hops flowers that are integral to the brewing process. Alpha acids are so important that hops varieties are rated according to how much they produce: The higher the percentage, the higher the potential bitterness. Be aware, though—the alpha acid percentages of homegrown hops can vary from year to year. For this reason, many homebrewers use their homegrown hops for aroma purposes only and use commercial hops for bittering.

The most overt result of these alpha acids is the bitter flavor they bring to beer. This quality adds to the distinctive flavor of beer and offsets the yeasty overtones of the malt. Alpha acids are released through boiling. Longer boil times free up more alpha acids, resulting in increasingly bitter brews. Hops that are used for bittering are usually added at the start of a boil that will last for at least sixty minutes. All hops release different levels of bitterness, aromas, and flavors depending on when they are added during the boil.

In addition, alpha acids have antibacterial properties that act to preserve beer. This antiseptic additive was a breakthrough for early brewers who knew the heartbreak of a spoiled cask. Before hops, beer was made with whatever herbs the brewmaster had on hand. Then one happy day, hops came to hand, and beer was redefined. Flavor was a big piece of the epiphany, but equally important were the microbe-killing capabilities of hops, which elevated the act of tapping a keg from potential disappointment to predicable joy.

In days past, hops with high levels of alpha acids were classed as bittering hops, while those with alpha acid ratios closer to those of its beta acids were termed aroma hops. This is no longer true. Breeding breakthroughs in the United States have simplified brewing with the development of dual-purpose hops, such as Simcoe, Amarillo, and Citra, which have high alpha acid levels but are also used for aroma. Adding hops after fermentation, also known as "dry hopping," is widely considered the most effective way to extract the delicate hop aroma. They also can be extracted during the boil, but if added too early, the aromatic properties are reduced. During fermentation and storage, the alpha acids gradually mellow and beta acids increase by oxidation. This process lies behind the distinctive taste and quality of lagers and aged beers. It's also why hoppy beers like India Pale Ales are best consumed fresh.

Hop Varieties

In the United States, hops are best adapted to Washington, Idaho, Oregon, California, and other regions with mild temperatures and low humidity. This does not, however, rule out home-grown hops in other parts of the country. At one time or another, hops have been grown in every state in the nation. In ultra-chilly New England and the Midwest, the sticky southeast, and the baking southwest, hops can be produced with a bit of extra planning and preparation.

Hops are hardy in USDA Zones 5 to 9, but winter hardiness is just one consideration. Know your region and the challenges hops plants may face. Does high humidity encourage diseases such as powdery mildew that can kill some varieties to their crowns? Do harsh, prolonged winters mean seeking out particularly hardy hops? Recommendations from local growers can go a long way toward your success. In the end, trial and error is your surest means to a productive hop yard.

The most challenging conditions for hops are hot, dry locales below the 35th parallel. (Yes, Phoenix, I'm talking about you.) But dedication can win the day. If you prepare a deep planting bed with lots of compost, are willing to experiment, and give your hops lots of shade and water (especially in July and August), you have a chance of overcoming your climate. In

Tettnang hops on the bine

these punishing regions, bittering hops may be a better bet than aromatics. In this case you (the hobby grower) actually have an advantage over commercial growers in that you can afford to coddle your few plants.

Hops Varieties

Brewer's Gold: bittering (high alpha) hop. Black currant, citrus. Vigorous. High yield. Susceptible to all hop diseases. Midseason harvest. OK for Midwest.

Bullion: bittering (high alpha) hop. Blackberry. Vigorous. High yield. Somewhat disease resistant. Late season harvest.

Cascade: aroma hop. Spicy, floral, citrus. Gives flavor and aroma to American light lagers, American-style pale ales, and many other brews. Vigorous. High yield. Tolerant of verticillium wilt and downy mildew but susceptible to aphids. Midseason harvest. OK for Midwest and New England. A good choice for the Southeast.

Centennial: bittering (alpha acid) hop. Floral, citrus. Good in pale ales. Vigorous. Moderate yield. Moderately resistant to downy mildew and verticillium wilt. Midseason harvest. OK for Midwest and New England.

Chinook: bittering (high alpha) hop. Very bitter. Grapefruit, pine. High yield. Good storage stability. Midseason harvest. OK for Midwest. Somewhat tolerant of downy mildew and highly tolerant of insects, so a good choice for the Southeast. Also a good try for the Southwest.

Crystal: aroma hop. Mild, slightly spicy. Citrus. Vigorous. High yield. Cone tolerance to downy mildew.

East Kent Goldings: aroma hop. Spicy, floral, honey. Low yield. Susceptible to downy and powdery mildew and wilt. OK for Midwest.

Eroica: bittering (high alpha) hop. Vigorous. High yields. Resistant to insects and disease. Midseason harvest.

Fuggle(s): aroma hop. Mild, grass, flowers. Traditional English ale hop. Low to moderate yield. Susceptible to verticillium wilt. Early harvest. OK for Midwest.

Galena: bittering (high alpha) hop. Black currant, grapefruit. Very bitter. Vigorous. High yield. Midseason harvest.

Glacier: dual-purpose (bittering and aroma) hop. Citrus, fruit. High yield. Susceptible to mildew. Midseason harvest. OK for Midwest and New England.

Golding: English ale aroma hop. Honey, earthy. Moderate yield. Susceptible to powdery and downy mildew. Early to midseason harvest. OK for New England.

Liberty: aroma hop. Spicy, citrus. Good for finishing German-style lagers. Moderate yield. Midseason harvest.

Magnum: bittering (alpha acid) hop. A clean bittering hop for ales and lagers. High yield. Tolerant of downy mildew and wilt; susceptible to powdery mildew. Good storage stability. Late harvest. Good option for the Southwest.

Mount Hood: aroma hop. Grapefruit, herbal. Vigorous. Moderate to high yield. Moderately disease resistant. Midseason harvest. OK for Midwest.

Newport: bittering (alpha acid) hop. Herbal, cedar. High yield. Resistant to powdery mildew, susceptible to downy mildew. OK for Midwest. Good for Pacific Northwest.

Northern Brewer: dual-purpose (bittering and aroma) hop. Spicy, resinous. Low to moderate yield. Tolerant of verticillium wilt, susceptible to powdery mildew. Midseason harvest. OK for Midwest and New England.

Nugget: bittering (high alpha) hop. Spicy, fruit. Good for light lagers. Vigorous. High yield. Disease resistant. Good storage stability. Midseason harvest. Popular in the Pacific Northwest. OK for Midwest and New England. A good choice for the Southeast.

Perle: bittering (alpha acid) hop. German lager hop. Citrus, cedar. Moderate yield. Early season harvest. OK for Midwest and New England.

Saaz: high quality aroma hop. Earthy, herbal, mildly floral. Weak grower. Low yield. Susceptible to downy mildew. Early season harvest. OK for New England.

Spalt Select: aroma hop. Herbal, floral. German lager and ale hop. Moderate yield. Tolerant of verticillium wilt, downy mildew. Late harvest. OK for Midwest. Good in Pacific Northwest.

Tettnang: aroma hop. Mild, spicy, herbal. Good for finishing German lagers. Moderately vigorous. Low yield. Tolerant of verticillium wilt. Early season harvest. Good for New England. Also a good try for the southwest.

Willamette: aroma hop. Mild spicy, grassy, black currant. American ale hop. Moderate yield. Tolerant of downy mildew, resistant to viruses. Midseason harvest. Grown almost exclusively in the Pacific Northwest. OK for Midwest and New England.

Zeus: bittering (alpha acid) hop. Black pepper, licorice. Vigorous. Hardy. High yield. Susceptible to downy and powdery mildew as well as aphids and mites. Mid- to late-season harvest. OK for Midwest and New England Closely related to Columbus and Tomahawk.

Hops in the Garden

At last, it's time to stop researching and start ordering your plants. Grab a cold one and find a source of hops plants that specializes in your region, often a local homebrew supplier. As you peruse their offerings keep this in mind: If your goal is to brew, start with at least three or four varieties. This not only helps ensure against crop failure but also gives you a range of flavors and alpha acid levels with which to experiment. Consider starting small with only a few plants. Four plants grown well will yield more and better hops than a whole row of neglected plants. A single well-grown hops plant can yield up to 6 pounds of fresh cones—plenty for most homebrewers. Although the initial crop may be disappointing, the harvest will increase in subsequent years.

Hops are sold as dormant rhizomes or potted plants. Hops in pots can be planted any time the soil can be worked, but may be hard to find in all seasons. Freshly dug rhizomes are offered only in early spring. If the rhizomes cannot be planted immediately upon arrival, wrap them in damp newspaper or damp sawdust and store in a cool place. The refrigerator works well if you have room in your crisper for large, lumpy roots.

When the time comes to plant your hops, siting is critical. For best production, hops want a site with full sun, good air circulation, soil that neither bakes nor bogs, and room for some form of trellising. Although a southern exposure is not absolutely necessary, less optimal exposures will result in smaller cones. Home growers should place one plant per hill with a manageable 3-foot spacing for plants of the same cultivar, or 5 feet between different varieties. The difference in spacing between like and unlike cultivars is one of order and identification: The closer the plants are, the more prone they are to hopeless tangling, in which case you'll never know which flowers you're picking. Tight spacing

Cascade hops on the bine

also lowers yields as the plants compete for light. Plant the rhizomes vertically with the buds facing up, or horizontally 1 to 2 inches deep after the threat of frost is past. Then stand back. At the peak of growth the plants can climb as much as a foot per day to a height of 20 feet or more. Bloom time is from mid- to late-summer. The plants should have at least ten to twelve productive years ahead.

Hops are climbers. Period. If not provided with support, they will find their own. Fences, trees, telephone poles, the neighbor's garage—nothing within 10 feet that can't run away is immune to their hoppy hug. That said, the bines' tiny, clinging hooks do not seem to damage the host, especially if the bines are removed each winter when they die back to the crown. To be safe, use caution in allowing hops to climb on porch supports or thin-barked trees.

Although hops will grow up practically anything they can wrap themselves around, for ease of picking consider growing your plants on strings or poles attached at the top to a strong horizontal

You can use a plastic pipe to support your hops.

wire. This system can be designed so that the wire can be unfastened and lowered for easy harvesting. Once the hops raise their heads a foot above the ground, coax them clockwise around their supports. As they grow, continue to steer them in the right direction as needed.

Fertilize first-year plants once a month with a balanced fertilizer such as 10-10-10. Follow the manufacturer's directions and apply to new plants from the first sight of new growth until early July. For established plants, apply the same fertilizer twice a month from emergence until flowering. A mulch of compost or manure applied once a year also helps the plants along. Water well during the growing season, especially for the first two years after your hops are planted, but don't let them sit in soggy soil. In swampy ground, plant the rhizomes in mounds raised 6 to 8 inches above ground level. If the site drains well, regular watering will fuel the bines' rocket-like growth and boost flower production.

Supporting Your Hops

A hops structure must, above all else, be sturdy. A fancy trellis or elegant arbor for hops is like a training bra for an underwear model: Too much stress and too little coverage. Forget lacy lattice; you need support that's up to the job. Although hops can be grown on a fence or other horizontal structure, the bines are equally adept at climbing straight up. Professional hops growers take advantage of this upstanding

nature with vertical trellis systems that allow maximum sun exposure to the plants as well as easier access to the cones. Hop yards are forests of 18-foot poles spaced 3 to 4 feet in rows 14 feet apart. Guy wires are attached at the top of the poles down each row. Upright wires are then placed along the row in Vs, starting at the ground and angling up to each side.

A system like this is certainly useful to a home hops growers, however, other supports can be considered. An old flagpole can be pressed into service, with three or more support lines spaced evenly around the base, beginning several feet away and attaching to the top. Even better, attach them to the pulley system so that the lines can be lowered at harvest time. If no flagpole is available, a similar system can be devised with a pole or 4x4 post. To avoid posts altogether, three or four support wires can be run from the ground and angled up to meet at the peak of the eaves of a building with a southern exposure; once again, a pulley makes bringing the bines down a simple job.

For most hops supports, poles or posts and some form of line is standard. Wire rope, aircraft cable, or heavy twine can be used. Hops bines will also happily scramble directly up the posts themselves or the narrow columns of a porch or pergola. Avoid lattice or chain link fencing. The bines will tangle hopelessly in anything they can, which will make the job of winter removal a bigger chore than it needs to be.

Developing New Hops

Not all hops cultivars are available to the homegrowing public. Many newly developed hops are protected by public and private breeders. Until the 1990s, most new varieties were released directly to the public. Now, however, it's common for private companies to keep their plants proprietary. That's not to say homegrowers will never have access to these varieties. Public breeders eventually bring their best creations to market, and the patents on privately held hops ultimately expire. And we have plenty to keep us busy in the meantime; while we wait for the new and novel, there are more than two dozen good and commonly available cultivars from which we can choose today.

Many modern hops cultivars have arisen, in part, from three key parents. Brewer's Gold (bittering), Fuggle (English aroma), and Hallertauer mittelfrueh (German aroma) are ancestors of most of the hops available today. Until recently, breeding programs focused on improved hops strains to replace existing ones. For example, a new variety may have improved resistance to spider mites, aphids, or powdery and downy mildew, or produce higher yields, but otherwise match the acid and alpha ratios of the plant it was designed to replace. This kind of linear breeding is purely practical; large brewers don't want to change their recipes, but hop growers benefit from productive plants that require minimal pest control.

The recent rise of the craft beer industry has led to a widening of some breeding programs to develop cultivars with unique chemical and aroma profiles. For professional hops breeders, the road from pollination to production is a long and expensive one. New releases can be in development for fourteen or more years. As the new varieties mature, hops breeders compile data on traits such as disease resistance, yield, age of maturity, coning habit, and many different chemical characteristics such as bittering acids and essential oil content. Hops take years to come to full fruition.

Harvesting Your Hops

You've been patient, sitting in the shade of your hops bower dreaming of worts and sparging. It's finally time to harvest those hops. How do you know? Beginning in August—as early as June in some regions—the cones should be checked regularly for ripeness. Indicators are a pleasant aroma, papery feel, and in some varieties, a change in color from green to yellow-green. Ripe cones feel dry and will leave behind a slightly sticky lupulin residue upon handling. The base of the bracts should be heavy with lupulin. To check the readiness of your crop, pull open a cone and look for this dark gold powder. Cones that are deep green, vaguely damp, and smell of hay are unripe; if these cones are pressed between the fingers they remain flattened while a ripe cone will spring back. If, on the other hand, cones are beginning to brown or smell unpleasantly strong, they have passed their prime.

Begin picking at the top of the plants where the cones are likely to be ready soonest. On a single plant the cones mature at different rates, so be prepared to harvest every few days until the cropping is complete, usually by the end of September. Wear long sleeves and gloves when harvesting to guard against the skin-raking bine spines. Remove the cones gently, with two hands, to avoid losing any of the precious lupulin. Use a picking basket or bag that can be slung over your shoulder or attached to your belt to keep your hands free. Once picked, the hops should never again be in direct sunlight—unless you prefer beer that has overtones of scared skunk.

The hops may be used immediately as green hops in a wet-hopped ale to commemorate the harvest, or dried for future use. In wet-hopped brews, hops fresh off the plant are added during the boil. This gives the resulting beer a fresh, springy flavor. The downside to this is the difficultly of knowing how much of the fresh hops to add. When fresh, hops have a much higher and more unpredictable moisture ratio than dried hops, and this makes it a bit tricky to judge the amount of green hops to be used in any recipe.

Drying Hops

Most brews rely on dried hops. Dried hops allow more precise control over lupulin amounts and, thus, the final result. For best quality, the hops must be properly dried as soon after harvest as possible to preserve their flavor and aroma. Hops can be dried via old-fashioned air drying or with mechanically assisted methods such as a fan, food dehydrator, oven, microwave, or even a homemade hop drier. Low and slow should be your mantra when drying hops. Heat drives off some of the aromatic complexity of the cones; too much heat will ruin them completely.

Two preservation methods, air and fan drying, avoid heat altogether. Air drying can be done in any warm, dark, dry room with good air circulation and enough space for the hops to be laid out in a single layer on anything from open paper bags to clean window screens. Fluff the cones every day or two and cover them with cheesecloth if dust is a concern. Your hops should be ready in three days to a week. When the lupulin is falling away and the stems snap when bent, the cones are sufficiently dry. If not thoroughly dried, the hops will mold and be unfit for brewing.

If you recognize the benefits of air drying, but crave a more dynamic and faster method than that of passive screen drying, put a fan beneath the screen. For air-dried hops in twenty-four hours, attach two screens together with a layer of hops in between. Bungee cord the result to a box fan. Keep watch over the hops in the final hours to keep them from becoming overly dry. If they dry too much, your lupulin can be gone with the wind.

Food dehydrators should be turned to the lowest temperature, around 95 degrees Fahrenheit, to dry hops. At this setting the cones may take up to three days to dry. To avoid having the house smell like a Bourbon Street bar on a Sunday morning, move the dehydrator to a garage or outbuilding. Ovens pose an even greater aroma risk, since the temperature may be higher and portability is impossible. Never dry hops at a temperature higher than 140 degrees Fahrenheit.

As with all heat-drying methods, micro-waves can reduce the delicate aromatic essence of hops. On the plus side, micro-waves can dry small batches of hops in only a few minutes. Dry the cones at fifty per-cent power. Stir every thirty seconds. When the hops are partially dried, remove them and give them time to finish drying. If they seem at all damp once they have cooled, return them to the microwave and continue the process. Once again, be prepared for a strong and potentially unpleasant smell.

Die-hard brewmasters can follow in the footsteps of the professionals by con-structing a hops dryer, known as an oast. Oasts, or hop kilns, are derived from tradi-tional English oast houses, which are the size of barns. Oasts rely on heated, circu-lating air for desiccation. A homemade, bee-hive-sized oast may not be as imposing as a three-story, furnace-heated architectural behemoth, but it is more appropriate to a backyard harvest. The best reason for con-sidering an oast is to deal with a very large home harvest consisting of pounds of hops.

Storing Hops

Dried cones should be sealed in airtight plastic bags. Pack the bags tightly and squeeze out all the air, or better yet, vacuum seal them. The hops can be refrigerated to be brewed within a week, or frozen for up to a year. When you're ready to fire up the kettle, the hops can be used directly from the freezer. Remove only the hops you need as thawing and refreezing can degrade the essential oils.

BARLEY (AND OTHER GRAINS)

If hops are the heart of beer, barley is its backbone. And for good reason: Barley is an alcohol alchemist. No other cereal grain contains as much of the fermenta-tion-friendly enzymes that break down

Cannabaceae

Hops, *Humulus lupulus*, is a member of the family Cannabaceae. Sound famil-iar? Cannabaceae, as is cannabis, as in marijuana. The two genera, *Humulus* and *Cannabis*, may not be instantly recognizable as relatives, but their flowering gives a clue to their kinship: both have resinous, imperfect flowers with male flowers on one plant, female flowers on another.

grains' stored starches into the sugars required for turning seeds into beer, whiskey, and other spirits. Barley's effect is so powerful that it acts as a catalyst to ferment other, less endowed, grains such as wheat, rice, rye, corn, oats, and millet.

Growing your own backyard barley may sound hardcore, but it's actually surprisingly easy. Barley, *Hordeum vulgare*, is a forgiving crop in northern climates. In addition, it is high yielding, matures early, and is widely adapted to all but the hottest and driest conditions. It is an annual crop that completes its entire life cycle within a year—usually sprouting in spring and seeding (and dying) in summer. If well sited and well tended, a 10x10-foot barley bed can deliver 5 to 15 pounds of grain, enough for one or two 5-gallon batches of all-grain brew, or a dozen or more batches of partial mash. A plot of this size will require ½ to ¾ pound of seed. The thicker the seeding, the less room there will be for weeds.

Perhaps the hardest part of growing barley is deciding which kind to grow. Two-row or six-row? Bearded or beardless? Hulled or hull-less? Really, the number of choices seems excessive.

"Beard" refers to the 3-inch bristle that extends from each seed. For the home-grower, the presence or absence of a beard is important mainly in that the bristles can be irritating to exposed hands and arms. They are also irritating to livestock and, on the plus side, deer. Hull-less is a misnomer as this kind of barley does have hulls, but

they drop away at threshing. The advantage of hull-less barley is that no special processing is needed to remove the hulls. This is good not only because the process is outside the ability of most backyard growers, but de-hulling hulled barley also removes much of the nutritional value of the grain.

Your choice between growing two-row or six-row barley may come down to where you live. The designations refer to the

Barley

Barley

BOTANICAL NAME: *Hordeum vulgare*
PLANT TYPE: annual grass
USDA ZONES: 3–9
HEIGHT: 2–3 feet
SOIL: well-draining, fertile loam
LIGHT: full sun
WATER: moist soil during germination, drier as the crop reaches harvest
GROWTH HABIT: grassy
PROPAGATE BY: seed. Some varieties are spring-planted and some are fall-planted. Use 4 pounds of seed per 1,000 square feet. Seed can be scattered by hand or by a broadcast seeder; rake into the top 1–2 inches. Keep soil moist during germination.

SPACING: Scatter seed across a fallow bed, or sow in the rows with twenty– twenty-five seeds per foot.
MONTHS TO BEARING: two months once growth begins in spring.
PRUNING: none
HARVEST: spring-sown barley matures in about 70 days; fall-planted barley ripens about 60 days after growth resumes in the spring. Reap when barley is dry. Cut, bundle, and shock (stack upright in bundles) to dry.
NOTES: Barley comes in two distinct types: two-row or six-row. The type you choose will depend, in part, on your region.
BEST USED IN: beer

number of rows of seeds on the seedhead: two-row barley has a row lined up neatly on either side of the center stem; six-row barley has—as billed—six rows that encircle the stem. Six-row barley is a deviation, a mutation of the ancestral two-row grain. Two-row is generally favored for malting and brewing, but it grows best where nights are cool. Two-row barley can handle hot days, but needs time to recover when the sun goes down. This makes it a good choice for coastal areas and much of the West, including Montana, Idaho, Washington, Colorado, Wyoming, and British Columbia.

Six-row barley thrives in climates that exhaust two-row varieties. It is the barley of choice in the Midwest and Mexico. Six-row barley can be grown in any region where two-row barley does well. To find out which barleys are best suited to your locale, inquire at local feed stores or your regional Cooperative Extension agency.

Although new hybrids have closed the gap in quality between two- and six-row barleys, differences in protein and enzyme levels still exist. Two-row barley contains more starch, making it ideal for brewing and distilling. However, it also has less

Talk Barley like a Pro

Bearded: a type of barley with a 3-inch, stiff, hair-like bristle extending from each grain of a barley seedhead. These bristles are properly called *awns*.

Flail: a threshing tool consisting of a wooden staff of about 3 feet long with a short heavy stick of about 2 feet swinging from it by a leather thong or metal rings

Hulled: barley with a hard, inedible hull surrounding each grain. Hull-less varieties actually have a loosely attached hull that falls away during threshing.

Malt: raw grain that has been soaked in water, allowed to germinate, and is ready for use in brewing and distilling

Mash: crushed malt steeped in hot water to form wort

Sheaf: a bundle of cereal grains harvested, but not yet threshed

Shock: a group of sheaves of grain placed on end with the seed-ends leaning against each other

Threshing: knocking grains off their stalks

Two-row vs. six-row: two different barley forms. Two-row barley has two rows of grains, one on either side of the stem. Six-row barley has six rows of grains encircling the stem. The two forms have slightly different ratios of proteins and enzymes that give them different roles in brewing. These ratios are also dependent on the specific barley cultivar.

Winnowing: pouring threshed grain back and forth between two containers repeatedly to allow the chaff to be blown away by wind, or a fan or blow dryer

Wort: unfermented beer. Technically, wort is the sweet liquid that comes from mash.

enzymatic power to break down those starches into simple, fermentable sugars during the mash phase. Six-row barleys traditionally contain less starch, but tend to yield more wort-soluble protein. Both types come in a range of beard and hull styles. Choose your barley based on your personal taste: every cultivar, whether six-row or two-row, infuses its own distinctive flavor, aroma, and color in beer.

Just when you thought the decisions were over, you must now decide whether to plant your barley in spring or fall. In regions where winter temperatures regularly dip below 20 degrees Fahrenheit, fall planting isn't an option, but in warmer zones, barley can be sown in September or October (at least four weeks before the first expected frost), overwintered, and grown on for an early summer harvest, usually sixty days after growth resumes. In all regions, spring plantings can be made in March or April for harvest approximately seventy days later. Certain types of barley are better adapted for either spring or fall planting.

Any good, fertile garden soil in a sunny site should produce a barley harvest with bragging rights. Seed can be scattered by hand or with a broadcast seeder. Rake the seed into the top inch of soil and water it in. Continue watering throughout summer until the plants begin to change color from green to tan. Allow the stalks to dry to honey brown. When the heads have turned golden, test to see if the grain can be easily pulled from the plant and has hardened to the point it can barely be dented by a fingernail. If so, it's time to harvest.

For the backyard farmer, a sharp hand sickle or a power trimmer with a string or blade attachment is the simplest tool for cutting the stalks. If using a hand sickle, hold the stalks with one hand and swing the blade with the other. Lay the stalks neatly along the ground as you go, with the heads in the same direction. If anyone asks, you can sound like a professional if you tell them you are making windrows. Likewise, when using a power trimmer, the goal is to cut the stalks so that they fall into windrows.

Now the barley will need to dry for approximately a week to ten days or until the grain can be shaken from the stalks. If you feel lucky, and live where rain is unlikely, you may simply let cut stalks lie. If wet weather looms, cover the stalks with a tarp. In damper areas, and for Norman Rockwell photo opportunities, stack your harvest into shocks. Simply tie the stalks into 12-inch bundles or sheaves. The sheaves are then stacked upright in teepee fashion in groups of twelve or more, forming the shocks. If hungry birds threaten your crop, cover the shock tops with fine netting or light cheesecloth.

Two-row vs. Six-row

Is six-row barley a sell-out to the man? North American barley farmers have long favored six-row barley because of its adaptability. In the mid- to late 1800s, the majority of US barley was grown where most cereal grains were grown, in the plains states. The hot days and nights of this region made six-row barley the logical choice. However, the six-row varieties available at the time had a downside: high soluble protein and enzyme levels that overpowered all-malt (all-barley) beer.

In other words, six-row barley had too much of the right stuff. To compensate for these overdeveloped levels of wort-soluble proteins, brewers added unmalted cereal adjuncts from corn and rice. The protein in corn and rice is largely insoluble and works to dilute the mix. These adjuncts can replace up to forty percent of a high-protein six-row grist.

August is a difficult time for barley. However, a late April planting can be allowed to dry to a cornucopia-worthy gold and left standing for effect. Or a later planting may be made with the express purpose of providing a green barley meadow to front your beer garden. Keep this crop well-watered during the hot months and there's a good chance you'll harvest grain by autumn.

This is where purists take issue. Although adjuncts were originally used out of necessity, the development of better balanced six-row barleys means that the inclusion of large amounts of adjunct is now mostly a function of tradition and profit. Most corporate brewers continue to include lower-priced adjunct in their beers. That is not, however, to say that high-protein six-rows don't have their place in the world of craft brewing—the super-charged diastatic power is useful for fermenting other specialty unmalted grains such as wheat, rye, and oats.

Threshing, Winnowing, and Storing

Threshing means separating the grain from the stalks. *Winnowing* is a technique for separating the chaff (bits of stalk and other debris) from the grain. Threshing requires muscle. Winnowing requires finesse. Both require patience. The two simplest methods by which the home gardener can thresh their barley are *flailing* and *bucketing.* Flailing is traditionally accomplished with a two-piece, wooden flail; however, for small harvests, the process works almost as well with a broom handle, plastic baseball bat, or other such stick (and removes the danger of concussing yourself with a badly aimed flail strike). Lay the sheaves on a dropcloth or bed sheet on a garage floor or asphalt driveway. Aim your threshing implement at the seedheads and whack away. Each sheaf should release a cup or two of barley.

Bucketing requires only a clean 5-gallon bucket or garbage can. In this version of

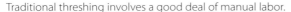

Traditional threshing involves a good deal of manual labor.

threshing, handfuls of harvested barley are held head-side down and beaten against the inside of the container. This method can take less time than flailing, but results in more chaff mixed with the grain, which will take more time in winnowing.

Winnowing uses wind, whether provided solely by nature or assisted by a fan or blow dryer. In winnowing, the grain is simply poured slowly back and forth between two containers. The heavier grain will fall into the receiving bucket while the wind blows away the chaff. Winnowing won't completely clean the grain; you'll need to pick out any debris heavier than chaff.

The easiest, most sure-fire way to store grain is in resealable bags in the freezer. Barley will maintain its quality for a year or more when frozen. In lieu of freezing, barley can be kept in airtight containers at a constant temperature of 60 degrees Fahrenheit or cooler in a dry, dark place such as a cellar. It can be kept this way for up to six months if the grains are put away absolutely dry. Any moisture in dry-stored barley will result in rancid seed and a quick ticket to the compost heap.

MALTING

Why would anyone *want* to go through the complicated home-malting process? For the same reasons you grow the ingredients and brew your own beer, of course: quality, sustainability, and your love of experimentation! Barley is the best grain to start with, though with a few tweaks to the process, other grains will work with the following method.

The good news is that, while it can be tricky, there are just a few steps to malting grains (and if you are an experienced gardener, you probably already know how to germinate seeds). The best time to malt barley is from late fall to winter—depending on your basement temperature—and when you have approximately one week to complete the process. You need a room that you can keep at 50°F (10°C) or cooler. The cool temperatures keep the grain from growing a green shoot. The same temperatures also prevent the grain from growing mold, fungi, and mildew.

You will need:

- Two clean 5-gallon plastic buckets, at least one with a lid
- One sieve bucket (a 5-gallon plastic bucket with ⅛-inch holes drilled into the bottom for drainage)
- Thermometer
- Scale that can weigh up to 20 pounds
- Aquarium pump, tubing, and air stone as well as enough tubing to connect the air stone at the bottom of a 5-gallon bucket to the pump sitting outside the bucket
- Food dehydrator (recommended) or oven
- A basement or other dark space that stays at 50°F (10°C) or cooler
- Baking sheets for oven or a food dehydrator

How to Malt

1. Clean the Grain

Weigh your barley for its dry weight before you clean it. This is an important first step so you will know when your grain is fully dry after steeping. Pour the grain into a 5-gallon bucket and fill with water. Stir the grain, and then let it sink to the bottom. The chaff, small bits of stems, and weed seeds will float to the top. Scoop out the floating debris. Stir again, and allow any debris to float while the barley sinks. Continue until you are satisfied that your grain is free of debris. Empty the cleaned barley into your sieve bucket and allow the water to drain completely.

2. Steep

The process of steeping barley takes seventy-two hours. Place your cleaned grain in the bottom of one of your 5-gallon buckets and cover with a ½ gallon of cool water at 50°F (10°C) or enough water to stand 2 inches above the grain while steeping. Allow this to stand for two hours. Pour the water and grain into the sieve bucket again and let the water drain completely.

Steep your grain for eight hours, then drain and let it sit for eight hours without water. Soak again for another eight hours, then check if the grain is plump, swollen, and showing white bulges. Timing is everything, so choose your eight-hour intervals for a time when you will be awake. We know your grains seem like your babies—you've nurtured them up to this point. Nevertheless, you can put these babies on a strict schedule that doesn't have you waking up in the middle of the night for a water change or drain.

Continue draining and changing the water every eight hours for the next seventy-two hours. Keep checking the grain for readiness to germinate. At the end of the steeping process, your grain will be plump and swollen. The tips of the barley grain show a whitish bulge from their emerging roots. Alternatively, you can use an aquarium pump with an air stone for aerating the water during the steep. With aeration, you can leave the grain soaking in the water and only need to drain and change the water every twenty-four hours.

3. Germinate

You are now entering the shady side of the process, which requires a dark room with the temperature set close to 50°F (10°C) during germination. This is important—a dark, cool room prevents the grain from growing a green shoot, which is great for growing more plants but destroys your malt.

Here is where you'll be grateful you invested in a pump and air stone. The air stone oxygenates the water below the sieve bucket full of germinating grain, which means you won't have to *turn* the grain.

The aerated water below the germinating grain keeps the area free of carbon dioxide buildup, which can smother the grain. Aerated water also helps the grain stay moist and prevents the germinating mass from becoming overheated in the middle.

Meticulously drain your grain as you have before in your sieve bucket. Pour a gallon of fresh water into the bottom of your other bucket and slip the air stone at the end of the tubing into the water. Connect the tubing to an aquarium pump sitting outside the bucket. Place your sieve bucket inside the bucket with water. Double-check that you haven't squashed the tubing to the air stone and reduced the air flow. The sieve bucket with barley should be above the water level so that the grain does not come in contact with the water.

Now wait. After approximately three days, rootlets will appear, and a shoot called an acrospire will bulge from the husk. When the acrospires grow ⅔ to ¾ the length of the barley, it's time to move on to the next step.

4. Couch

Remember how we didn't want carbon dioxide to build up around the germinating grains so as not to smother the seeds? Couching the malt means the carbon dioxide becomes our friend. We stop the acrospires in its tracks by denying the malt much-needed oxygen for growth. This is where the process becomes interesting—the grain starches turn into fermentable sugars. This is the easiest step in the process and takes one to three days to complete. Simply unplug the aquarium pump and seal the bucket with its lid.

Check the barley and turn the mass of grain once every twenty-four hours. We don't want the carbon dioxide to kill the malt, just to stop the growth. When the growth is stopped, it's time for the next step.

5. Kiln

Drying the malt is called *kilning*. This is the step where you remove the moisture you added during the process. You can use an oven or a food hydrator to kiln your malt, though since kilning requires low temperatures and a long time, an inexpensive food dehydrator is the more practical option.

First, spread your grain out on baking sheets. Then place the sheets in your dehydrator or oven and set it at, or close to, its lowest setting—you want a temperature between 90 to 125°F (31 to 50°C). If you are using your oven, keep the oven door partly open for air circulation during the kilning, and work with a partner to make sure it remains attended overnight. Most malt will require twenty-four to forty-eight hours to reach the desired moisture content. You will know your malt is dried properly when it weighs the same as when you weighed the grain before you began the malting process. The easiest way to check on your grain would be to remove ¼ or ⅛ the batch and weigh it.

At this point, the barley will still be pale. Since you're working with a small amount—typically less than you would need for a full batch of all-grain beer—you may want to go a step further and create a light or dark crystal malt. Either way, take a break, and transfer the barley to colanders to shake off any rootlets that are still attached.

To toast the malt, transfer the malt to your oven. Toasting the malt at 275°F for an hour should create light crystal malt. For medium crystal malt, toast at a higher temperature of 350°F for twenty to thirty minutes. For dark crystal malt, keep the heat at 350°F, but toast it longer than a half hour. You may need to stir the malt partway through to ensure even toasting. Also, keep a very close eye on the malt as it begins to darken.

Wheat and Spelt

BOTANICAL NAME: *Triticum aestivum* (wheat); *Triticum aestivum* subsp. *Spelta* (spelt)

PLANT TYPE: annual grass

USDA ZONES: 4-8

HEIGHT: 2–3 feet

SOIL: well-drained, fertile loam

LIGHT: full sun

WATER: moist soil during germination, drier as the crop reaches harvest. Too much water can cause wheat to fall over. For wheat, water two to three times during a dry summer. For spelt, water only during germination.

GROWTH HABIT: grassy

PROPAGATE BY: seed. Some varieties are spring planted and some are fall planted. Use 4 pounds of seed per 1,000 square feet. Seed can be scattered by hand or by a broadcast seeder; rake into the top 1–2 inches. Keep soil moist during germination.

SPACING: Scatter seed across a fallow bed,or sow in the rows with twenty–twenty-five seeds per foot.

MONTHS TO BEARING: four, once spring growth begins

PRUNING: none

HARVEST: Reap when wheat is dry. Cut, bundle, and shock (stack upright in bundles) to dry.

NOTES: Spelt is an ancient species of wheat. Wheat and spelt do best with a cool, moist growing season followed by warm, dry weather for ripening.

BEST USED IN: beer

After you plant the seeds, cover the crop with a frost cloth to keep the birds from gobbling up your seeds.

Testing Wheat For Readiness

When the grains start to bend down, it's time to test for readiness. The only equipment you need for this is your mouth. Pop the grain into your mouth and give it a chew. If it feels soft, it's not time. Keep testing daily until you bite down and the grain crunches. Time to harvest!

Harvesting a small plot of wheat can be as simple a pruning the heads off of the stems, or using scythe or sickle. Pruning the grain heads is time consuming, but you just have to prune and toss the wheat heads into a basket. If you use a scythe, adding a cradle to the tool makes it a lot easier when it gathers the grain as you cut. Otherwise you have to spend a lot of time picking up the wheat arranging it for binding into sheaths. With this method, you will want to put your handfuls of cut wheat pointing in the same direction.

Bundle sheaves by taking a handful of wheat and binding it together with twine, then place them in a dry place to cure, protected from animal marauders. The grain is cured when it shatters easily.

Thresh the grain to separate the chaff from the grain by beating each sheave inside a large metal container, such as a clean trash can. To winnow the wheat, you can either pick a breezy day or set up a fan. Pour small batches of wheat from one container to another, letting the wind or fan blow the chaff away. Repeat as necessary until the chaff is removed.

Spelt is an increasingly popular grain, and some brewers have started experimenting with it as well. It has a distinctive nutty flavor. Like wheat, you can grow two crops of spelt per year. Even better, spelt generally requires very little—it often needs no fertilizer and gives a better yield than wheat on poor soil. If you live in the right zone, simply plant it in early spring on freshly-raked dry soil, water it well, and keep it free of weeds afterward. Toward the end of summer, your spelt will be ready for harvest. You can then harvest, thresh, and winnow as you would for barley. Keep it cool and dry until you're ready to malt it.

Spelt can also be malted as easily as barley. Since wheat is difficult to malt and not recommended for the homebrewer, spelt is an alternative and has a nuttier taste than its newer prodigy. Like wheat, you can grow two crops a year. However, spelt needs no fertilizer and can give a better yield on poorer soil. The crop needs lit-

Lupulin

Bitter, resinous lupulin is nature's way of encouraging passing herbivores to spit out the hops cone and go find a nice carrot. But, as with the pepper plant's mouth-burning capsaicin, the defense backfired. Humans found virtue in the seemingly inedible. Although lupulin may not have kept the hops flowers on the plant, it was wildly successful in propagating the species. Under the patronage of humans, hops have been widely hybridized, nurtured, and spread around the globe. Pretty smart for a plant.

tle care other than planting in early spring on freshly raked dry soil and watering it in, and keeping it weed-free afterwards. Towards the end of summer, spelt is ready for harvest. Harvest, thresh, and winnow as you would with barley. Keep the grain cool and dry until you are ready to use it.

Malting spelt is done similarly to barley. Let 5 pounds of spelt soak in a 5-gallon bucket of water for twelve hours. Drain and rinse with cold water and drain again. Germinate the grain in a sieve bucket and turn it as you would for barley to keep the grain from molding. Rinse and drain with cool water every eight hours until the spelt germinates in two or more days. Kiln the spelt the same way you would barley, when the acrospires—or initial sprout arising from the germinating seed—is the same length as the grain.

Hops for Decoration

Dried hops cones are as decorative as they are useful. If you can spare a few from the brewing tank, use hops cones to decorate the table or as part of an arrangement to celebrate Oktoberfest. Or put them in the spotlight of an even more important occasion with hops corsages and boutonnieres. The preserved cones are fragile, but flexible and very aromatic. To avoid the chance of clothes-staining lupulin, dry the cones before they are fully ripe. If the cones won't be used immediately, store them in a resealable plastic bag in the refrigerator. As opposed to hops to be used for brewing, pack these hops lightly in the bags to preserve their shape. Hops bines can also be used for decoration. Each year in late summer, fresh and dried bines are draped on beams and bars across England in hops-producing regions. To make your own dried bine garland—complete with cones—place cut bines in a dark, warm place for several weeks. It is helpful but not necessary to provide them good air circulation with an electric fan.

CHAPTER 3

Other Beer Ingredients
Brewing Herbs and Fermentables

O ne of the many benefits of brewing your own beer is that you can make much more unique recipes than most commercial breweries. Jalepeño beer? Go for it. Cucumber lager? Why not? Gruit ale? Sign me up. Kale ale? Let's not go crazy. Still, for a passionate brewer, beer can be so much more than hops and barley. Herbs, fruits, seeds, vegetables, and trees—whether grown or gathered—add entirely new dimensions to homebrew.

Ancient brewmasters knew this well. For much of its estimated 12,000-year history—ever since humans began to collect and soak cereal crops—beer was made from a wide selection of botanical ingredients. The recipes seem to have been limited only by season, region, and blatant toxicity. In this chapter, you'll find everything from the classic brewing herbs to citrus fruits, many of which can easily be grown out back or on the balcony.

GETTING INTO GRUIT

Before hops became part of the standard ingredients list for brewing beer, gruit (also spelled grut or gruyt) was a common style of ale. The word may sound strange, but like many brewing terms it actually originated in Germany: *Gruitbier* translates roughly to "herb beer" and is pronounced *groot beer*.

Gruit has many variations and interpretations, as the term describes any beer that uses a combination of herb ingredients for bittering and flavoring beer. Historically, these recipes were often closely guarded secrets, but common herbs included sweet gale (*Myrica gale*), marsh rosemary (*Rhododendron tomentosum*), heather (*Calluna vulgaris*), ground ivy (*Glechoma hederacea*), horehound (*Marrubium vulgare*), yarrow (*Achillea millefolium*), juniper (*Juniperus*), or lavender (*Lavandula*) in a variety of ways.

The literature of those times spoke about the extreme inebriating qualities when these herbs fermented. Today's cautions include watching the amount you drink, as some of these ingredients make the beer more intoxicating. If you don't want to be an excited, wide-awake drunk—

easy does it. Some gruit beers can excite you as they inebriate you (whereas hops behave more like a sedative).

Since this book is primarily about gardening, we're only recommending general quantities when it comes to brewing for the herbs in this chapter. In other words, we're the gardening experts, not the brewing experts! When in doubt, use even less than the ballpark amount. And if the initial flavor is still too sharp, letting the beer age by waiting a few months after bottling will help mellow the flavor. For specific recipes and usage amounts, we recommend reading a complementary book, such as *Sacred Herbal and Healing Beers.*

Warning: Always research your specific plants and recipes before brewing.

Many herbs and some other plants can be poisonous and others can act as an abortifacient (an abortive agent). If you are pregnant, you probably aren't consuming beer to begin with, but if you do have the occasional sip, make it a non-herbal variety.

Gruit Plants

If making medieval-style beer intrigues you enough to want to grow and brew an herb or three, the good news is that most of them grow with ease in a wide variety of gardens. Provided there is enough sunshine on the space, even an apartment balcony garden using large pots can grow many of the herbs you need to brew a batch of Gruitbier.

Yarrow

Yarrow is the most widely used medicinal herb species from Europe, and the plant has naturalized in the northern temperate regions around the globe. So perhaps it's not surprising that it also ended up as an ingredient in ales! Historically speaking, the herb has a long association with Homo sapiens, and our distant and extinct relatives. A story about yarrow pollen found in 60,000-year-old Neanderthal burial caves makes for an interesting topic of conversation while sitting around a backyard fire with friends sipping a brew made from this herb. Its botanical name, *Achillea*, refers to a 3,000-year-old story from the Trojan War, in which Achilles used yarrow on wounds to stop the bleeding.

Yarrow

BOTANICAL NAME: *Achillea millefolium*

PLANT TYPE: perennial herb

USDA ZONES: 3–10

HEIGHT: 2–3 feet

SOIL: well-drained to moist; tolerates poor, dry soil

LIGHT: full sun

WATER: low to moderate

GROWTH HABIT: clumps spread 18–36 inches by underground runners.

PROPAGATE BY: division

SPACING: 2–4 feet

YEARS TO BEARING: one year

PRUNING: Cut old stems to ground during dormant season. After a few growing seasons, yarrow needs rejuvenating. By the third year, the herb becomes a thick mat of leaves with reduced flowering. Divide and replant, or smother the old plot with a layer of four sheets or more of newspaper, covered in thick mulch. A few plants left on the edge will take over with fresh, new growth and lots of flowers.

HARVEST: The first of two to three harvests is taken when the leaves and flowers first show in late spring. Pick flowers and only healthy green leaves. After the first and second harvests, cut the yarrow all the way down to encourage new growth. After the third harvest, let the herb grow. Use the flowers and leaves fresh when possible. To dry, cut the stems and flowers in the morning, tie the stems together, and hang upside down in a well-ventilated, dark, dry space.

NOTES: Choose a white-flowered strain for beer. Yarrow can be invasive in the garden. Good for container culture.

BEST USED IN: gruit beer

Brewers frequently used the highly aromatic flowers and leaves of the common yarrow plant (*Achillea millefolium*) as a substitute for hops to give beer a satisfying floral and herbal quality.

Linnaeus, the father of botany, originally gave yarrow the name *Galentara*, which meant "causing madness," perhaps in reference to its role in alcoholic beverages.

You'll find the herb grows with ease in the garden; actually, it grows with wild abandon. For a brewer's first foray into grow-your-own, this would be a great starter plant. However, choose a white flowering strain for your beer and leave the colorful flowering strains for the ornamental garden. Yarrow can be grown in pots or in the ground, as the herb can handle a variety of growing conditions. Wet or dry, shade or sun, open or deep woods, meadows, prairies, or your lawn, you'll find yarrow is a plant that adapts. However, the optimum conditions for harvesting are full sun and moist, fertile, well-drained soil. The plant likes to romp beyond its bounds, so either grow the herb in a container or somewhere between a foundation and a sidewalk. In two years, for example, yarrow spaced four feet apart will be one solid mass of plant! Although it would be a difficult task to harvest each plant separately, it would be interesting to plant a mint and yarrow opposite each other in one bed and let them duke it out to see which one comes out a winner. Both are invasive plants, so just plant with caution and stand back!

After a few growing seasons, yarrow will need rejuvenating. The herb will continue to live and grow, but it settles in with a thick mat of leaves and doesn't flower as prolifically as it does when invading new land. You can smother an old plot of yarrow by covering with a thick layer of newspaper, covered in thick mulch. Leave a few plants on the edge and let it invade a new area or divide the plants and start over in another plot. You can also alternate where you allow it to invade, so it is always on the move conquering new ground and happily flowering in the process.

The first harvest of the leaves and flowers begins when the flowers first start to show in late spring to early summer. Adjust the timing of your harvest accordingly to accommodate your climate and weather patterns. You can usually get two to three harvests in a growing season. When harvesting leaves, choose only those that are fresh and green; discard any shaded out yellow to black leaves underneath. After the first and second harvests, cut the yarrow all the way down to encourage new green growth. After the third harvest, let the herb grow on its merry way.

Time your brew schedule with the harvest of the flowers and leaves to use them fresh. For later use, cut the stems and flowers in the morning; tie the stems together and hang upside down in a well-ventilated, dark, dry space; and let them dry at temperatures that don't exceed 95°F. Store your dried herbs in airtight containers.

Because of yarrow's antiseptic, antimicrobial, and antibacterial properties, it acts as a preservative in ales—much like hops. The strong, astringent-tasting flowers or leaves should be used sparingly in the brewing process or the plant will overpower the beer. One ounce of dried yarrow or up to two ounces of fresh flower tops is a good starting point, or start with some tried and true recipes before you venture out with your own creative concoctions.

Marsh Rosemary

A beautiful member of the heath family, marsh rosemary can be found growing wild in swamps and other wet climates in Europe, Asia, and North America. It may not look much like the rosemary you know, but you'll find out why it has "rosemary" in its name when you bruise the aromatic leaves and get a whiff of its potent rosemary-like scent! Most rhododendrons have large leaves, but this shrub's foliage looks needle-like (also similar to the real rosemary herb). Marsh rosemary shrubs (also known as wild rosemary, marsh tea, and crystal tea) have long been used for medicinal purposes and are thought to have narcotic properties that make beer more intoxicating. Because the plant contains the toxin Ledel, use caution with this plant and do not heat the leaves in a closed container.

Marsh Rosemary (Wild Rosemary)

BOTANICAL NAME: *Rhododendron tomentosum*

PLANT TYPE: evergreen dwarf shrub

USDA ZONES: 2–6

HEIGHT: 1–4 feet

SOIL: peaty, boggy

LIGHT: full sun to part shade

WATER: regular

GROWTH HABIT: low-growing, shrubby

PROPAGATE BY: difficult to propagate

SPACING: 24–36 inches

YEARS TO BEARING: two years to harvest

PRUNING: Pinch prune to keep desired shape.

HARVEST: At the time of flowering, harvest only new leaves growing on new stems without bark. Dry the leaves using a drying rack in a ventilated area.

NOTES: This is a plant with a strong, resin-like aroma. A member of the heath family, marsh rosemary is thought to have narcotic properties that make beer more intoxicating. Although the plant contains toxic substances, it has long been used to flavor beer, instead of or along with hops.

BEST USED IN: gruit beer

Once classified as *Ledum palustre*, marsh rosemary possesses analgesic, anti-inflammatory, antimicrobial, antiviral, antifungal, and insecticidal potential. Sounds appetizing! Recent research shows wild rosemary as having promising results as having antidiabetic, antioxidant, and anticancer properties, a testament to the powerful nature of this plant. Still, as a beer ingredient, marsh rosemary provides no guarantee that your beer is medicinal.

You'll find marsh rosemary plants are handsome additions to a landscape and are especially useful if areas of your property have boggy, wet, or peaty soils in full sun or partial to full shade. Once planted, the slow-growing dwarf shrub will be a long-term resident of your garden, as a handsome, bushy, spreading evergreen. If you don't have a wet, boggy area, keep the plant well irrigated. Supplement with a deep watering of about an inch of water each week during the growing season if nature doesn't provide it.

The plant dislikes lime, which raises the pH level of the soil, so grow the shrub away from areas such as lawns if you sweeten your soil to a higher pH level. Because the plant grows better in a symbiotic relationship to some fungal organisms in the soil, encourage microbial organisms by not using fungicides, herbicides, or insecticides near this plant. The shrub grows 1 to 4 feet tall and is covered with dull, dark evergreen leaves.

Attractive corymbs of white flowers appear from May to June. At the time of flowering, harvest only the new leaves growing on the new stems without bark. Grow a patch of these beautiful shrubs and harvest a few leaves from each plant. Because the plant is a painfully slow-growing shrub, it is important not to over-harvest any individual plant.

Harvest the leaves and dry them on a drying rack in well-ventilated, dark, dry space, at temperatures that don't exceed 95 degrees Fahrenheit. Store your dried herbs in airtight containers.

An alternative brewing ingredient to the marsh rosemary is Labrador tea (*Rhododendron groenlandicum*). However, marsh rosemary is said to have the best flavoring of the two. Use up to ¼ ounce dried marsh rosemary to start with when mixed with other herbs or find specific gruitbier recipes that call for this ingredient.

English Lavender

The English lavenders (*Lavandula angustifolia*) are used for complex bitterness in beer, much like heathers (*Calluna vulgaris*). Also used in gin and liqueurs, English lavender imparts a milder taste than other species of lavender.

In a Mediterranean-style garden, these shrub-like herbs are stunning with their silver-gray leaves and purple-blue flowers. *Lavandula angustifolia* 'Tucker's Early',

'Munstead', or 'Hidcote' are good choices to grow for your brew. *L.* 'Tucker's Early' is a compact erect plant that is one of the earliest blooming English lavenders, yet other cultivars work as well.

English lavenders prefer the Mediterranean-like, dry summers of the West Coast. However, if you live in an area with steamy, hot summers, you can substitute *L.* x *intermedia* cultivars that can handle more humidity such as *L.* 'Grosso' or 'Gros Bleu' that impart a sharper flavor. Other lavender species outside of *L. angustifolia* or x *intermedia* cultivars are not suitable for consumption as they are mildly toxic.

You don't need to fertilize lavenders. After the initial flower harvest when the flowers begin to fade, shear your plants back to keep them compact and tidy. Prune above the wood where there are leaves. The stems won't regrow if you cut into the wood.

Bundles of lavender ready for drying

English Lavender

BOTANICAL NAME: *Lavandula angustifolia*

PLANT TYPE: shrubby evergreen herb

USDA ZONES: 5–8

HEIGHT: 2–3 feet with a spread equal to or twice that, depending on cultivar

SOIL: well-drained, neutral pH

LIGHT: full sun

WATER: low

GROWTH HABIT: low, shrubby perennial

PROPAGATE BY: softwood cuttings in summer

SPACING: 1–2 feet, depending on cultivar

YEARS TO BEARING: two years until harvest

PRUNING: Shear down half of new growth each year after bloom or the following year before growth begins. The stems won't regrow if you cut into the old wood. Lavender plants grow woody and gangly over time; replace them when they become unsightly.

HARVEST: Pick the flowers when the blooms are just opening up. Tie

stems together and hang upside down to dry in a cool, dry location.

NOTES: English lavender lends a milder taste to beverages than other lavender species. *L. angustifolia* 'Tucker's Early', 'Munstead', or 'Hidcote' are good choices for brewing, yet other cultivars work as well.

BEST USED IN: gruit beer, gin, liqueurs

Lavender plants will grow woody and gangly over time, so many people replace their lavender plants after five years. You can propagate your plants by taking softwood cuttings in summer.

Harvest the flowers when the blooms are just opening up. Tie the stems together and hang upside down in a well-ventilated, dark, dry space, and let them dry at temperatures that don't exceed 95 degrees Fahrenheit. Store your dried herbs in airtight containers.

A good starting point is about ½ ounce of fresh or dried flowers added towards the end of the boil. You can also use the flowers as you would hops for dry hopping.

Heather

In the nineteenth century, Robert Louis Stevenson wrote a poem about the Scottish version of the legend of heather ale. It's a gruesome tale of a battle set in pre-Roman times, where a Pict (thought to be pre-Celtic people) gave up his life and that of his son, rather than reveal the secret of a heather ale recipe to the Scots. As many legends go, the truth is bent, or stretched beyond repair. There was no genocide of the Picts; many of their descendants live in parts of Scotland today.

The Irish have their own version of the legend, only the last man was a Viking in the last clash with the Irish. Rather than give up the recipe for *bheóir Lochlannach* (which translates as Viking beer), the Viking chooses death. Still, even without the recipe, Viking beer most likely was a heather-flavored mead or cider. The idea of brewing a tasty, legendary ale is intriguing. With many heather ale recipes readily available, you won't need to threaten anyone for a secret one!

If you wish to grow *Calluna vulgaris* for such a purpose, a dozen or more plants will be necessary to harvest enough for your brew. It takes up to 3 quarts of fresh-picked flowering heather tips for a 5-gallon batch

Purple heather plants are as beautiful as they are functional in a multi-purpose garden.

Heather

BOTANICAL NAME: *Calluna vulgaris*
PLANT TYPE: low evergreen shrubs
USDA ZONES: 5–7
HEIGHT: 8–24 inches
SOIL: well-drained, slightly acidic
LIGHT: full sun
WATER: low, but best with summer irrigation
GROWTH HABIT: low-growing mounds or spreading mats
PROPAGATE BY: semi-ripe cuttings in midsummer, or sow seed as soon as ripe and place in a cold frame for the winter
SPACING: Heathers vary widely in spread, so space the plants about as far apart as the plant's mature width as listed on the nursery tag.
YEARS TO BEARING: once plants are large enough to provide ample tips

PRUNING: If heather is used for brewing, harvest by shearing when the flowers are fresh in late summer, which will double as the plants' yearly pruning. Otherwise, shear lightly after flowers fade to keep the plant from becoming straggly. Avoid pruning into woody barren branches, which won't produce new growth.
HARVEST: Shear fresh flowers in late summer.
NOTES: At least a dozen plants are required to harvest enough heather tips for brewing. Good rock garden or raised bed plants. The heather tips are added to the boil and sometimes the sprigs are dry-hopped in some recipes.
BEST USED IN: gruit beer, cider

of beer. Fortunately, the plants are small enough that you can easily grow many of them in your garden. Some homebrew supply stores carry dried heather too, but heather is beautiful in the garden and easily harvested, so why not grow your own?

Your best bet is to grow plants in full sun, in well-drained, slightly acidic soil. However, heather grows well in rock gardens too. The plants can tolerate dry soil, but will need irrigation during a prolonged summer drought. Shear lightly after flowers fade to keep the plant from becoming straggly. Also, avoid pruning into woody barren branches or the stem won't regrow. Propagate by semi-ripe cuttings in midsummer or sow seed as soon as ripe and place in a cold frame for the winter.

It is preferable to pick the fresh flowers in late summer; however, you can dry the stems by bundling and hanging them upside down in a cool, dry, dark place. The heather tips can be added to the boil. Some recipes also suggest dropping sprigs in beer after primary fermentation (similar to dry hopping).

Sweet Gale

Sweet gale literally gets around, growing circumpolar in the northern latitudes throughout the world. In North America, its native haunts are throughout Canada and Alaska but it is equally at home in Oregon, Washington, the Great Lakes states, New England states, and even dips down into North Carolina and Tennessee.

Before hops appeared, sweet gale may have been the most prominent gruit ingredient in Britain. Beer enthusiasts have a difference of opinion on the claim that both sweet gale and marsh rosemary are brewed in the same historic gruit recipes since the shrubs grow in different regions in Europe. It may be that the flavoring that went into a regional gruit would have used the shrub native to where the brew was developed, not both as some claim. Considered the true gale, sweet gale appears to be the preferred flavoring of the two as evidenced by some of the common names of the time for marsh rosemary—false gale and pig's gale. There are many common names for *Myrica gale* such as bog myrtle, Dutch myrtle, meadow fern, and English myrtle, yet sweet gale is neither a myrtle nor a fern!

The shrubs are dioecious, meaning they are either male or female; however, some individuals can change from male to female, and back again. If you wish to propagate your plants from seed, you will need at least one male and one female plant, and hope they are not fickle enough to change to the same gender in any given year. The

Sweet Gale

BOTANICAL NAME: *Myrica gale*

PLANT TYPE: low, deciduous shrub

USDA ZONES: 2–9

HEIGHT: 4–6 feet

SOIL: damp, peaty, acidic

LIGHT: full sun to light shade

WATER: constant; suitable for bogs

GROWTH HABIT: shrubby

PROPAGATE BY: softwood cuttings, layering, seed

SPACING: 4–6 feet

YEARS TO BEARING: when large enough to give up the required number of leaves

PRUNING: Pinch to keep dense; use harvesting as pruning.

HARVEST: At the time of flowering, harvest only new leaves growing on new stems. Don't over-harvest an individual plant. Dry the leaves using a drying rack in a ventilated area.

NOTES: Sweet gale is a soil-improving nitrogen-fixer.

BEST USED IN: gruit beer, gale ale

male catkins emit pollen that is carried by the wind to pollinate the female catkins. (If you have pollen allergies you won't want to aggravate the problem by growing this in your garden.)

Sweet gale is a bushy, 2- to 4-foot deciduous shrub. The nitrogen-fixing abilities of nodules in the sweet gale's roots and its symbiotic relationship with a fungus called *Frankia* mean it can grow in nitrogen-poor soil. As with marsh rosemary, do not use any fungicides near the shrubs. In its native habitat, sweet gale grows in perpetually wet conditions next to lakes and streams, and in fens, bogs, and swamps. You can grow this with the moisture-loving marsh rosemary, as they require similar conditions. Since both shrubs have similar flavorings, choose one if you are short on garden space. Neither shrub can tolerate lime and prefer acidic, peaty soil.

Covered in glands, the glossy, blue-green leaves emit a pleasing, sweet-smelling, resinous fragrance. Harvest the leaves and dry them on a drying rack in well-ventilated, dark, dry space; and let them dry at temperatures that don't exceed 95 degrees Fahrenheit. Store your dried herbs in airtight containers.

To make a gale ale use ½ ounce of dried sweet gale leaves and mix with other herbs such as lavender and rosemary.

Warning: the plant is an abortifacient; if you are pregnant do not consume beer made with this ingredient.

Ground Ivy (Glechoma hederacea)

The name *Glechoma* comes for the Greek word *glechon*, which means mint-like plant. As with its other relatives in the mint family, this running groundcover comes with a warning: ground ivy loves to spread like many other mints. This species has enough common titles to name a couple of litters of kittens, such as creeping Charlie, Lizzie-run-up-the-hedge, alehoof, herbe St. Jean, cats foot, field balm, runaway-Robin, or Gill-over-the-ground.

In spite of its well-earned reputation for being an opportunist in the lawn and gar-

ground ivy

den, the plant is worth growing as an ingredient for your brews. As it creeps along the ground, its wiry stems grow roots wherever it goes. If you don't mind the plant taking over your garden, grow it in the ground. However, you can keep it from taking over by growing it in a container, where it does well. A variegated form is often sold as an ornamental groundcover, or as a trailing plant in hanging baskets and containers. For brewing, we recommend sticking to the species *Glechoma hederacea*.

A bitter, aromatic herb, ground ivy was the most common plant used for clarifying and preserving beer until the sixteenth century when hops took over its job. Some will grow it to add to salads, cook in soups and stews, or brew in mixed herbal teas.

The low-growing herb grows best in full to part sun in medium soil with good drainage. Space plants 5 inches apart. Harvest leaves from spring to autumn, but the best harvest is in May when it is at its peak flavor. Use the leaves fresh (2 to 4 ounces) or dry (1 ounce) per dried 5 gallons. Simply tie the stems together and hang upside down in a well-ventilated, dark, dry space; and let them dry at temperatures that don't exceed 95 degrees Fahrenheit. Store your dried herbs in airtight containers.

Warning: the plant is an abortifacient. If you are pregnant do not consume beer made with this ingredient.

Ground Ivy (Creeping Charlie)

BOTANICAL NAME: *Glechoma hederacea*

PLANT TYPE: low-growing, evergreen, perennial herb

USDA ZONES: 3–10

HEIGHT: under 6 inches

SOIL: well-drained

LIGHT: sun to shade

WATER: moist to dry

GROWTH HABIT: a creeper with wiry stems (stolons) that root where they touch the ground

PROPAGATE BY: Dig up plants from self-rooting stems

SPACING: 3–6 inches

YEARS TO BEARING: As soon as the plants produce enough leaves

PRUNING: Dig out and remove as needed.

HARVEST: Harvest leaves from spring to autumn. Use the leaves fresh or dried. Dry them at temperatures that don't exceed 95 degrees Fahrenheit.

NOTES: A variegated form is often sold as an ornamental groundcover or used in hanging baskets and containers. Can be invasive, especially in moist shade. To be on the safe side, grow in a pot.

BEST USED IN: gruit beer

Horehound

This is an herb used in gruitbier that is associated with magic and herbal remedies; it is an effective expectorant in cough medicines, for example. You could claim your horehound ale to be medicinal. However, horehound most likely loses much of its medicinal properties, such as soothing sore throats and reducing inflammation, once you brew with it.

This is another plant with many common names, the most common being horehound (sometimes spelled hoarhound), common horehound, and houndsbane. In Germany, the herb is called *Berghopfen*, which translates to "mountain hops." Horehound has been traced all the way back to the Egyptians who called the herb seed of Horus, eye of the star, and bull's blood. All three names sound like good ale titles!

The herbaceous perennial plant is native to Europe, Northern Africa, and Asia and has naturalized in the United States and Canada where it was introduced. Like the dandelion, it has made itself at home in the new world and can be found in almost

Horehound

BOTANICAL NAME: *Marrubium vulgare*

PLANT TYPE: perennial herb

USDA ZONES: 4–9

HEIGHT: 1½–2½ feet

SOIL: can survive poor soil; does well in rich soil as well

LIGHT: full sun

WATER: low to moderate

GROWTH HABIT: erect stems

PROPAGATE BY: seed; division; basal cuttings (3- to 4-inch long shoots with some of the underground stem attached)

SPACING: 12 inches

YEARS TO BEARING: as soon as plants produce enough leaves

PRUNING: Cut back old stems before new growth starts. The second year, prune back the plants by half—preferably when harvesting.

HARVEST: Before the plants begin to flower, harvest the leaves in the morning after the dew dries. The first year, cut no more than one-third of the growth. In subsequent years, the plants can be cut back to 4 inches from the ground. Use the leaves fresh or, for later use, dry them at temperatures that don't exceed 95 degrees Fahrenheit. Once the leaves are dried, their musky fragrance disappears.

NOTES: can be invasive. Remove flowers to prevent self-seeding. To be safe, confine it to a pot.

BEST USED IN: gruit beer

every state and province. The genus belongs in the mint family and bears a resemblance to mint with its crinkled leaves that are covered in white hair—it has a woolly gray appearance. The white flowers bloom from late spring into August and are arranged in whorls in the axis of a pair of opposite leaves on the square stems.

The plant adapts well to poor, dry soils, yet will thrive in good garden soil too. Plant with other herbs, or if you think the nettle-like plant looks too weedy for your garden, plant them in an out-of-the-way area. Like many members of the mint family, it may be invasive. It self-seeds, so you want to deadhead the flowers religiously, not letting any seeds form.

You can start your plants from seed by sowing them in either spring or late summer in pots. For late summer to fall sowing, put your pots in a protected cold frame over winter. Germination can be slow, so don't give up on them. Keep the soil moist during the germination process. Take 3- to 4-inch-long basal cuttings in late spring by harvesting the shoots with some of the underground stem attached. Keep the soil moist while the plants establish their root systems. You can start them in pots and plant them in the garden in summer. The best way to propagate more plants is to divide clumps in spring and replant immediately where you want them to grow.

When your plants begin to flower, harvest your herbs in the morning after the dew dries. As you pick, you will notice the musky fragrance. Use the leaves fresh or, for later use, tie the stems together and hang upside down in a well-ventilated, dark, dry space; and let them dry at temperatures that don't exceed 95 degrees Fahrenheit. Store your dried herbs in airtight containers. Once the leaves are dried, the fragrance disappears.

When brewing a horehound ale or traditional English ale with this bitter herb, don't use more than 2 ounces of fresh leaves or 1 ounce dried per 5 gallons.

Juniper

Common juniper is a low-growing, or upright, evergreen bush native to high northern latitudes around the entire globe. Here in North America it grows in alpine meadows, the plains, and as an understory plant for wooded areas of native conifers and quaking aspen trees (*Populus tremuloides*). Historically, Native Americans used juniper medicinally as a tonic for an assortment of ailments and used the branches as a smudge to repel insects. They believed that the branches brought good luck, protected you from disease, kept evil spirits away, and protected you from lightning and thunder.

The berries are still used today in potpourris, but more importantly as an ingredient for beer! (After brewing, feel free to toss any of your unused berries into the potpourri dish.) Both berries and boughs have been used during the brewing process and for filtering out grains and hops in the bottom of brew vessels, and juniper can be found in recipes for porters, stouts, and doppelbocks. (It was also used as one of the ingredients in a recipe for Goat Scrotum Ale—an unfortunate but memorable title.)

Common juniper goes by other names such as dwarf juniper, prostrate juniper, mountain common juniper, old-field common juniper, and ground juniper. Some good selections for the garden and for brewing include *Juniperus communis* var. *charlottensis*, *J. communis* var. *megistocarpa*, *J. communis* var. *depressa*, and *J. communis* var. *montana*. The last two varieties rarely grow taller than 1 to 3 feet.

The low-growing conifer is a popular groundcover for the garden because it grows on a variety of soils including stony or sandy soils. Junipers will grow in acidic to alkaline pH soil, but in the garden, you want maximum growth potential; it will grow its best in a sandy-loam to loamy soil in full sun.

J. communis is not recommended for homes located in dangerous fire zones,

Juniper

BOTANICAL NAME: *Juniperus communis*

PLANT TYPE: coniferous evergreen shrub

USDA ZONES: 3–8

HEIGHT: shrubs, 1–13 feet; trees to 30 feet

SOIL: tolerates poor soil, but best in a sandy loam

LIGHT: full sun to shade

WATER: dry to moist; can tolerate drought

GROWTH HABIT: low-growing or upright shrub or small columnar tree. Common juniper has many cultivars with a wide variety of growth habits.

PROPAGATE BY: cuttings

SPACING: dependent on cultivar

YEARS TO BEARING: once plants are large enough to harvest foliage. Cones, three years.

PRUNING: rarely needs pruning. Snip back individual stems to shape. When harvesting or pruning boughs, don't cut below the green needles or the branch won't regrow.

HARVEST: Ripe, plump berries should be gathered in autumn and dried slowly in the shade to retain the oil.

NOTES: Any juniper berry brew should be avoided by mothers-to-be. *J. communis* is not recommended for homes located in dangerous fire zones because its resinous foliage is highly flammable.

BEST USED IN: beer, gin

where fire-resistant landscapes are a necessity. The plant is highly flammable with resinous foliage that burns more intensely than many other herbaceous plants.

These plants rarely need to be pruned, unless they're placed where their width overcomes a sidewalk. Site your plants carefully in the first place and you won't have deal with plants outgrowing their space. When harvesting or pruning boughs, it is important not to cut below the green needles. Only prune where needles appear along the stem, otherwise the branch will not regrow.

A small amount of potent juniper berries goes a long way, if you don't want your beer to end up tasting like gin. One ounce of berries per 5 gallons added the last ten to fifteen minutes of the boil is a commonly advised starting point.

Warning: women who are pregnant or anyone with kidney disease should not ingest juniper berries, or anything made with them. Because gin (also called mother's ruin) is flavored with juniper berries, it has the ability to cause miscarriage, so any juniper berry brew should be avoided by mothers-to-be.

THE CUCURBITS

Cucumber saison, pumpkin ale, and watermelon wheat are just some of the beers that are possible thanks to the cucurbits family of plants.

Melons can be problematic, but pumpkins, squash, and cucumbers are not hard to grow in most climates, except for the shorter-season ones. The biggest complaint we hear about growing cucurbits is the powdery mildew they can get later in the season. The fungus is unruly once the weather turns cool and humid! When the mildew sets in, the plant's ripening process halts permanently. However, researchers in South America found that a foliar spray of milk controls powdery mildew on squash, cucumber, melon, and pumpkin plants. Just mix a spray of one part nonfat milk to nine parts water and spray weekly to reduce the disease-causing organisms (*Sphaerotheca fuliginea*). Not only does the spray keep the fungus in check, it fertilizes the plants at the same time. This is an inexpensive, organic, and easy way to protect and grow healthy cucurbits.

Cucumber beetles and squash borers can also attack the plants. The best line of defense is using a row cover on the plants until they start flowering. Take the covers off at that point, or the bees won't be able to pollinate the plants.

OK, enough about the problems . . . on to the plants! Cucurbit vines produce male and female flowers. Plenty of beginner gardeners ask why their flowers are not producing fruit early in the season. It's because the first flowers to open are typically the male ones. (You can tell the difference between the male and the female, as the female flowers have a swollen ovary at their base that looks like a baby fruit.)

Once the female flowers open, if you are not seeing any fruit develop you may need to channel your inner bee and do some hand pollinating. Cut a male flower off the vine and pull off the petals. You will see the pollen on the stamens. Simply touch the pollen to the female flower's sticky pistil. Also, if you don't have enough pollinators in your garden, you may want to learn how to attract them to your plot of land. If you are spraying with pesticides, you may be killing off the beneficial insects who pollinate plants, keep pests at bay, or both.

Pumpkin and Squash

When fall moves in, we love pumpkin flavor in almost everything. Of course, beer is no exception! The best pumpkins grown as an edible are actually squashes though. In fact, canned pumpkin—the kind you might use for cooking or for brewing—isn't really pumpkin, but squash. Why they call it pumpkin is a mystery. Perhaps canned squash doesn't sound as appetizing, even if the fruit is sweeter and purees into a beautiful consistency for pie. Still, fresh is always best, for baking and for brewing!

Squash is not a tough vine to grow and harvest, but it does take up a lot of room in the garden. One solution is to plant a "three sisters" combination in your allotted space, which is an ingenious, ancient Native American way to grow corn, squash, and pole beans in the same plot. The beans are a legume that fixes nitrogen into the soil, while the corn stands tall and feels entitled to steal all the nutrients. The beans wind their way up the corn, while the pumpkins and squash wander around at the base, effectively covering the bare ground and shading out the weeds. Or, you can plant just pumpkin and just grow some vines to cover ground. The large leaves are actually quite handsome in a tropical looking way.

Try to give your vines full sun, although a little bit of shade won't harm them. Give them a lot of room to grow—overcrowding them reduces fruit yield. Space hills 4 to 5 feet apart and plant four to six seeds per hill. After the seedlings emerge, thin the sprouts to the most vigorous plant in each mound. Don't pull the weaker ones out; cut them to ground level with your pruners to prevent disturbing the roots of the chosen one. When sowing the seeds or transplanting, add one cupful of a complete organic fertilizer to each hill and fertilize every two to three weeks after that.

Harvest your squash after the first light frost when the vine wilts, and before a hard frost. When you cut the squash from the wilted vine, leave 2 inches of stem on the fruit. Leave the fruit out in the sun to cure for a few days. Protect your fruit from a hard frost during the curing time.

The amount of pumpkin or squash you'll use in a pumpkin beer recipe varies, but you'll want to think in pounds rather than ounces if you want any flavor to come through. If you want your beer to really smack of pumpkin, there are two tips. First, consider using some pumpkin pie spices in your beer recipe as well—while they will mask the gourd flavor somewhat, and they also remind the drinker of pumpkin desserts when she smells or tastes

Pumpkin and Squash

BOTANICAL NAME: *Cucurbita* spp

PLANT TYPE: vining annual vegetables

USDA ZONES: 3–9

HEIGHT: 2½ feet trailing, or 6–8 feet if trellised

SOIL: rich and well-drained, amended with compost

LIGHT: full sun

WATER: regular; at least an inch of water a week

GROWTH HABIT: trailing vines

PROPAGATE BY: seed planted directly in the ground. If the growing season is short, seed can be started in peat pots and planted out when the soil is warm.

SPACING: 18–36 inches. Plant several seeds in one spot—or hill—and thin to the best seedling.

MONTHS TO BEARING: four to five

PRUNING: none

HARVEST: A pumpkin or squash is ripe when its skin turns the solid color of its type (typically orange) and the rind resists puncture when pressed with a thumbnail. Harvest after the first light frost when the vine wilts, but before a hard frost. Leave a 2-inch stem on the fruit. Leave your squash in the sun to cure for a few days. To prepare for fermenting, cut the cucurbit in half, scoop out the seeds and pulp, and place skin-side-up on a roasting pan. Don't cover. Add ¼ inch of water and bake for approximately an hour at 325 degrees Fahrenheit. For fermenting, don't overbake the squash to mush. When cooled, remove the rind.

NOTES: Pumpkins and squash are heavy feeders. Apply fish emulsion every month, beginning when plants are 8 inches tall. The first flowers are male, and so will not become fruit. The best pumpkin for eating is squash; even commercially canned pumpkin is actually a mix of winter squash. Be warned that pumpkins make a difficult brew, but the flavor of the ale is worth the effort. The best flavor comes from putting the cooked fruit in the mash, rather than in the boil. The trade-off for good flavor is the amount of time it takes to run-off the wort to the kettle.

BEST USED IN: craft beers; notably, pumpkin ale

the beer. The second tip is to chop up and bake your squash on a cookie sheet before brewing with it. Some caramelized (but not burnt) pieces are optional but can provide additional depth of flavor. When the fruit cools, mash well by hand or puree the fruit in a blender. If you plan to freeze your squash, divide it up in amounts called for in your recipes to brew at a later date, or to steal for home-grown pumpkin pies.

Flavor-wise, there are many excellent cultivars to consider. The first is Rouge vif d'Etampes *(cucurbita maxima)*. Commonly called the Cinderella pumpkin, this plant is actually a squash. Its orange-red skin, deep lobes and low profile make it a favorite in the kitchen. I always plant this old French heirloom in my pumpkin patch. Average weight per pumpkin is 15 pounds and yield is two pumpkins per plant. 'Sunshine Kabocha' is a beautiful and delicious winter squash with space-saving, 6- to 8-foot vines. 'Waltham' butternut squash is a classic, culinary favorite that is an important part of the "pumpkin" in canned pumpkin. Any type of acorn squash will have bright, creamy flesh and a nutty-sweet flavor, but a compact early such as 'Early Acorn Hybrid' produces 4–6 large fruits that have a chance of ripening even in short-summer areas; they are also good for container culture.

Atlantic Dill, Big Moon, or Prizewinner (*Cucurbita maxima*): These are the most likely contenders for the Great Pumpkin that poor Linus van Pelt, of the Peanuts gang patiently waits for in the pumpkin patch every Halloween. These squashes are the ones that most avid growers cultivate as giants to enter into the biggest pumpkin contests. However, you don't want to grow one to its maximum size for your pumpkin ale. Let these vines set many pumpkins to keep the size from becoming too large to move or fit in the oven come roasting time.

Butternut or acorn squash (*Cucurbita moschata*): While definitely not a pumpkin, butternut and acorn squash are delicious and some of the easiest winter squashes to bring from harvest to puree. These provide great flavoring for pies and have a smooth flesh, so grow some for Thanksgiving deserts and put the rest into your favorite beer recipe.

Cucumbers

It appears that cucumber beer is catching on. We have not tried one, but have read reports that it is a refreshing, invigorating drink. If you have an overabundance of cucumbers, this will be a fun summer brew to try.

This popular warm-season crop has about the same cultivation requirements as pumpkins and squash. Unlike their cousins, cucumbers don't need the same amount of real estate to grow in the garden, and you can grow the vine up a trellis system or grow the bush types on the ground or in a 5-gallon container if you are short on space. If you trellis your cucumber plants, you can space them closer together at 6 inches apart. Growing them on the ground, you will need to space your plants 18 to 36 inches apart depending on

Cucumber

BOTANICAL NAME: *Cucumis sativus*

PLANT TYPE: vining annual vegetable

USDA ZONES: 4–11

HEIGHT: 2 feet trailing, or can be trellised to 6–8 feet

SOIL: rich and well-drained, amended with compost

LIGHT: full sun

WATER: regular. Apply a gallon per week once the fruit sets.

GROWTH HABIT: trailing vines

PROPAGATE BY: seed. Sow seed in the ground two weeks after the average date of last frost. To get an earlier start, seed indoors about three weeks before you will transplant them in the ground. They like bottom heat from a heat mat.

SPACING: 18–36 inches. Plant several seeds in one spot and thin to the best seedling.

MONTHS TO BEARING: two to three

PRUNING: none

HARVEST: Harvest slicing cucumbers when they about 6 to 8 inches, picklers at 4 to 6 inches. Pick when they are uniformly green and before they begin to yellow. Keep the fruit picked or the vines will stop producing.

NOTES: Bush-type cucumbers can be grown in containers. Cucumbers are heavy feeders. Apply fish emulsion every month, beginning when plants are 8 inches tall. Add cucumber to beer as a secondary to help retain the fresh, crisp flavor. Put it in sooner and you lose the flavor.

BEST USED IN: Craft beers

the variety. The seed packet will tell you the best spacing for a particular variety.

The day before sowing cucumber seeds, presoak the seeds in water overnight. Sow two to three seeds per hill. When they germinate choose the best seedling and prune or pinch out (don't pull) the other plants to the ground. Cucumbers grow quickly and are heavy feeders, so throw in a handful of a complete organic fertilizer at planting time and give them monthly feedings after they sprout. Plants grown in containers will need a weekly feeding. Keep the plants consistently watered or you will end up with hollow, bitter fruit. When hot weather arrives, containers can dry out quickly and you may need to water daily. If you grow your vines vertically, keep training them up their trellis.

You might want to try the somewhat sweeter tasting lemon cucumber as an alternative to the green one for your beer. Either way, cucumber is an ingredient you will want to add after primary fermentation, as the flavor is quite delicate.

Melons

We confess: We have melon envy. The mild climate in the Pacific Northwest makes it difficult to grow melons without jumping through a lot of hoops (or under them when growing the plants in a hoop house). There are avid gardeners in the region though that love the challenge and grow melons every year, with a variety of results. If you live in a warmer part of the country, you

Lemon cucumber

can grow them a lot more easily. Like other cucurbits, they are heavy feeders and need room to grow, and you'll want to fertilize them monthly for best results.

For the amount of space they need, melons have slow-growing vines and yields are not very high. You must be a serious gardener and brewer to want to grow melons to put in your brew. However, you will earn the bragging rights for this crop.

Keep your melon patch weeded so the vines won't have to compete for nutrients.

Melon

BOTANICAL NAME: *Cucumis melo* (musk-melon varieties); *Citrullus lanatus* (watermelon)

PLANT TYPE: vining annual vegetable

USDA ZONES: 4–11

HEIGHT: 1 to 2 feet trailing. Trellised muskmelons can reach 6 to 8 feet.

SOIL: rich and well-drained, amended with compost

LIGHT: full sun, three months of heat

WATER: consistent; cut back on water as the fruit begins to ripen. Use soaker hoses or drip irrigation to keep water off the leaves

GROWTH HABIT: trailing vines

PROPAGATE BY: seed. Plant when the ground temperature is above 70°F. Presoak seed to increase germination. Plant seeds two weeks after average last frost date. Or transplant from starts sown six weeks earlier in peat pots and grown in a greenhouse or under lights. Don't disturb the roots at planting out.

SPACING: 1 foot apart if grown on a trellis, or 3 to 4 feet apart if allowed to sprawl. Sow several seeds per spot and thin to the best 2 to 3 seedlings.

MONTHS TO BEARING: three

PRUNING: none

HARVEST: For muskmelons, hold the stem in one hand, and give the fruit a twist with your other hand. When the melon is ripe, it slips off the stem easily. If the melon doesn't slip off, leave it for a few more days and try again. For watermelon, pick when the curly tendril closest to the fruit turns brown.

NOTES: Muskmelons can be trellised. Amend the planting area with 4 to 6 inches of compost or well-rotted manure. Like other cucurbits they are heavy feeders and need room to grow. Apply fish emulsion every month, beginning when plants are 8 inches tall. Choose watermelon varieties that are *fusarium* and bacterial wilt resistant.

BEST USED IN: craft beers

Consistent watering is a must for best flavor. You want your soil to be humus-rich, because when the melons begin to ripen you will need to cut back on the amount of water you add to the soil. The humus in the soil retains moisture while keeping the soil from becoming too soggy, which in turn will give you better tasting melons.

Muskmelons (cantaloupe being one of them) can be trellised if you are short on space. Presoak your seeds to aid in germinating. The seeds will not germinate if soil temperatures are below 65 degrees Fahrenheit. You can prewarm the soil by putting a plastic tunnel over it. Plant seeds two weeks after the average last frost date for your area in hills 1 foot apart if you grow them vertically, or plant them 3 to 4 feet apart and let them wander around on the ground. Alternatively, you can transplant starts that were sown six weeks earlier and grown in a greenhouse or under lights. The starts resent transplanting so plant them carefully with little disturbance of their roots. Cover with a plastic tunnel or cage to aid in keeping the soil warm. To harvest your melons, hold the stem in one hand, and give the fruit a twist with your other hand. When the melon is ripe, it slips off the stem easily. If the melon doesn't slip off, leave it for a few more days and try again.

Watermelon is another beer-friendly melon. But if you can get your watermelon harvest into your beer before it is eaten, you have more willpower than we do! Two incurable diseases that are prevalent in watermelon are fusarium and bacterial wilt, so choose disease-resistant varieties. As with other summer melons, watermelons need warm soil to germinate and warm weather to ripen. Plant seeds two weeks after the average last frost date for your area. Alternatively, you can transplant starts that were sown six weeks earlier and grown in a greenhouse or under lights. The starts resent transplanting so plant them carefully with little disturbance of their roots. Space planting hills

about 6 feet apart and sow four to six seeds per hill. Thin seedlings down to the healthiest three. You can also grow in rows with one plant every 2 to 4 feet. Cover with plastic tunnels to keep the soil warm if the weather is still on the cool side, or nighttime temperatures are dipping down below 50 degrees Fahrenheit.

It is important to give your watermelon plants regular deep watering the first month. After the vine starts to set fruit, cut back on watering so as not to dilute the fruit sugars and make the fruit less flavorful. You can tell the watermelon is ripe when the curly tendril closest to the fruit turns brown.

A FEW FINAL IDEAS

You've seen the range of plants that made gruit ale popular for centuries, thought about watermelons, and maybe even planted your squash or cucumber. What else could there possibly be for your beer? Well, there are plenty of herbs, for one. However, since we have most of these conveniently gathered in the liqueur chapter (see page 177), simply flip over there to learn more about the herb of your choice.

On the few pages that follow, we'll run through the last few ingredients for specialty beers on our lists—consider them a few final ideas for your creative homebrew garden.

Spruce

Spruce needle beer is a spring tradition that dates back centuries, mainly because making the beer with fresh spring needles brings a refreshing, unsweetened cola-like taste to

Spruce

BOTANICAL NAME: *Picea* species

PLANT TYPE: coniferous evergreen trees

USDA ZONES: 2–8 depending on species; prefer cool, mild summers

HEIGHT: 4–150 feet depending on species and type. Dwarf varieties available, which can be grown in a container on the deck or as part of the landscape.

SOIL: moderately moist, sandy, acidic, well-drained soil

LIGHT: full sun to part shade

WATER: low to moderate

GROWTH HABIT: pyramidal to conical, dwarf varieties available

PROPAGATE BY: cuttings, grafting

SPACING: dependent on cultivar

YEARS TO BEARING: one to two years

PRUNING: little needed; light pinching or shearing to maintain tidy shape

HARVEST: You don't need to harvest many needles for the brew, but keep in mind to always use the freshest new needles.

NOTES: Spruces are low-maintenance landscape plants. They do not perform well in areas of high heat and humidity. Fresh, young, spring needles are used to give beer a refreshing, unsweetened cola flavor. Mature needles are not used because they taste like turpentine. Spruce essence is available in brew shops, but doesn't compare to the taste of fresh needles. Many species of spruce can be used for flavoring.

BEST USED IN: beer

beer. (If you harvest the needles later in the season, your beer will smell more as though it's made for thinning paints!)

Kumdis Island Spruce Beer recipe calls for the needles from the tallest tree of the species—Sitka spruce (*Picea sitchensis*). Kumdis Island, where the recipe traditionally comes from, is in the Charlotte Islands on the west coast of Canada. The recipe utilizes the needles from the indigenous spruces that grow into giants in the island's mild maritime climate.

The new shoots of many spruces are a great source of vitamin C, which helps stabilize the finished beer as well. Historically, the British Royal Navy added the needles to their ship-brewed beer to treat scurvy.

Harvesting fresh needles from the wild can mean climbing towering trees, but it can also be as simple as hunting down young trees. In fact, if you're looking for a small tree to plant, your yard can easily accommodate a dwarf version of a recommended spruce in a container on the deck, or as part of the landscape—and you can use the first growth of fresh, new needles.

If you live in an area where you can grow a spruce, they are wonderful conifer trees for the home landscape. For the Pacific Northwest, a dwarf spruce such as Sitka spruce (*Picea* 'Papoose') would be the best choice. The conical Alberta spruce (*Picea glauca* 'Conica') or blue spruce (*Picea pungens*) would be better suited for many colder regions. The Colorado blue spruce,

Picea pungens 'Fat Albert' is worth growing for the name alone but is also an outstanding garden specimen. Choose the best-recommended variety for the region you live in. The trees will reward you as a beautiful specimen in your landscape and provide some fresh needles for your spring brew.

Most recipes use a very small quantity of needles—not more than 4 or 5 ounces. Even though you don't need to harvest many needles for the brew, keep in mind to always use the freshest new ones. Even a small amount of old spruce can ruin a beer. Also, if you've ever bought spruce essence in homebrew shops, keep in mind that fresh spruce is very different. While the bottled stuff can come off like a pine-cleaning product, fresh spruce in small amounts can make a light bodied beer with a cola-like taste—and it could become a new favorite spring brew.

Lemon flowers

Citrus

Living in the Northwest, we are citrus challenged, as are many parts of the rest of the country. Unless you are living in Arizona, Florida, and parts of California—or equally warm areas of your continent where you can grow them in the ground—growing your citrus in a container will be necessary.

To grow citrus successfully in pots, choose varieties grafted onto dwarf rootstock, which are suitable for confinement in a container. Some cultivars will do well growing outside in cold climates when brought into a greenhouse for the colder months. Other varieties can transition to houseplant status when the weather turns cold outside. Even a balcony garden can accommodate a dwarf citrus tree as long as it is growing in full sun.

As in any other gardening endeavor, choose the best varieties for your specific region to increase your chances of success. Often a plant does not grow well because it was the wrong variety for the climate, soil type, and so on. In warm zones where you can grow your trees and shrubs in the ground, check with your County Extension agent to find out the best varieties suitable for your county. Another source for information on good cultivars for the home garden is a local citrus garden club. Some clubs may even have local plant sales or exhibits of the best cultivars to grow. The big box stores may sell citrus, but the plants may be more suitable for another region and not yours, so buy your stock plants from reputable, locally-owned garden nurseries that choose the best varieties for your area. Invest in good quality stock that will give you fruit for many years to come.

Most brewers use just the peels of the fruit as an ingredient, so even if you don't get a great crop of fruit, the peels may still be usable. For use in liqueurs and cocktails, it is more critical to grow good fruit for fresh, fragrant garnishes.

Left: Meyer lemon fruit
Right: A small orange plant fits nicely on a balcony

Citrus

BOTANICAL NAME: *Citrus × limon* (lemon), *C. latifolia* and *C. aurantifolia* (lime), *C. × sinensis* (orange), *C. × paradis* (grapefruit)

PLANT TYPE: broadleaf evergreen trees

USDA ZONES: 9–11 in the ground. Farther north in a container with a greenhouse or sunroom.

HEIGHT: to 50 feet depending on variety; dwarf 3–10 feet

SOIL: Citrus is adaptable to a range of soil types. Does poorly in salty soil.

LIGHT: full sun. The bark of citrus is thin and must be protected from intense sun.

WATER: regular. Citrus trees have relatively shallow root systems. Water well over the entire root zone. Established trees will tolerate some drought but won't produce high-quality fruit.

GROWTH HABIT: dense

PROPAGATE BY: grafting, budding, cuttings

SPACING: dependent on type and variety

YEARS TO BEARING: three to five

PRUNING: Remove suckers coming from below the graft. Remove any shoots that extend beyond the general shape of the tree. Thin overly dense foliage to promote air circulation and light penetration. Citrus can be trained as shrubs or hedges, or limbed up into a tree. If the trunk is exposed, it will need protection from sunburn. Many citrus types have formidable thorns; protect yourself with gloves and goggles.

HARVEST: Oranges, lemons, and grapefruit should all be completely free of green coloration. These fruit don't ripen off the tree. Limes don't change color, so judge ripeness by size and season.

NOTES: Where you live determines which types and varieties of citrus you can grow. Even gardeners out of citrus-friendly climates can grow containerized Improved Meyer lemons, bringing them into the house or greenhouse for the winter. In fact, the farther you live from the equator, the better the quality of your citrus fruit— until you hit frost. In order from least- to most-resilient to frost: lime, lemon, grapefruit, orange, Meyer lemon. Fertilize trees in January through March. Citrus in containers should be fed January through June. A fertilizer with micronutrients is important, especially for container-grown trees. Containerized Meyer lemons are more than a novelty; they're handsome, relatively undemanding plants (keep them well fed and watered; check for spider mites and scale) that smell heavenly in bloom and, with age, can produce pounds of delicious, slightly sweet lemons.

BEST USED IN: liqueur, wine, infusions for cocktails, beer

Corn

Corn has long been thought of as the domain of the macro brewers and tasteless, light beer. However, with the resurgence of craft lagers, and thanks to other creative recipes, corn is regaining some popularity.

For the home gardener, corn may be a satisfying crop to grow in the backyard. Although you can grow corn in rows, for best pollination it is better to grow the crop using the square foot gardening method. Basically, grow your corn in 4x4-foot squares, spacing your seeds so that one corn plant occupies one 12x12-inch square, which equals sixteen plants per 4x4-foot square. Each plant will produce one to two ears of corn, so calculate how many plants you will need before you plan how many squares or rows to prepare for planting.

If you have a long, warm growing season you can plant early-, mid-, and late-season varieties. If your season is short, spring remains cold well into June, or your summers are typically mild, early season varieties are a good choice. If you are concerned about genetically engineered corn, or the chemicals used to grow them, growing your own is one way to control what goes in to the food you consume. Select organically grown seeds and grow them yourself. Seed houses such as Johnny's Selected Seeds in Maine or Territorial Seed Company in Oregon sell a good selection of corn varieties suitable for cooler climate- or short-season growing and offer GEO-free (mistakenly called GMO) seed.

A corn crop isn't something you can grow easily on a balcony or in a container. Corn needs room to grow and will grab all the nutrients in the soil to help it grow tall and strong. You need a stand of corn for good pollination or you will wind up with kernel-less ears of corn. It is possible to grow your corn in a pot; however, given the cost of potting soil versus yield, it may not be worth the trouble for a dubious outcome.

Growing the same crops yearly means rotating them so that you renew the soil. Follow corn with a cover crop such as crimson clover that will add nutrients back to the soil after the heavy-feeding corn crop strips most of it away. Or use the sustainable, "three sisters" method of growing corn, squash and beans, mentioned in the cucurbit section on see page 79.

Corn

BOTANICAL NAME: *Zea mays*
subsp. *mays*

PLANT TYPE: vegetable

USDA ZONES: 4–8

HEIGHT: 4–10 feet

SOIL: warm (65 to 85 degrees
Fahrenheit) loam, amended
with compost and fertilizer

LIGHT: full sun

WATER: moderate to high. Water regularly and deeply. Soaker hoses make
the job simpler and more efficient.

GROWTH HABIT: large, leafy stalks

PROPAGATE BY: seed. Corn is best direct-
seeded, but can be sown in peat pots
at the last average date of frost and
planted out once the seedlings are
4 inches high.

SPACING: Sow in clusters of four to six
seeds every 12 inches in rows 24–30
inches apart; thin to the strongest
seedling when 4–5 inches high.

MONTHS TO HARVEST: three

PRUNING: none

HARVEST: Each corn stalk produces
two ears, occasionally three. To test
for ripeness, pull back the tip of the
husk until a few kernels are exposed
and pierce one with a thumbnail.
If it's plump and milky, it's ready
to pick. Corn of the same variety
planted on the same date bears over
a relatively short season, usually
less than two weeks. Store harvested
ears tightly wrapped in the refrigerator until you have gathered enough
for your recipe.

NOTES: Corn is wind pollinated, so
planting in large blocks (for example:
four rows, 10 feet long) gives better
pollination—and fuller ears—than
planting in one or two long rows.
Planting in cold, wet soil can cause
corn seed to rot. To prevent this,
use a soil thermometer to test that
the soil has warmed to at least 65
degrees Fahrenheit, and soak seed in
warm water (68 degrees Fahrenheit)
for about eight hours before planting,
which speeds germination. Keep soil
moist until the seed has sprouted
in about four to five days. Corn is a
heavy feeder, so fertilize twice during
the growing season with fish emulsion; once when the stalks are knee-
high and again when the tassels
start to show. A mulch of compost
or rotted manure is helpful to keep
corn's shallow roots moist and to
discourage weeds.

BEST USED IN: liqueurs, beer

Sugar Beets

Sugar beet roots look more like parsnips on steroids than your typical table beet and can grow as large as a basketball. You will want to pick them at a smaller size, however, when the sucrose is at its peak. The plant part we use is actually the taproot of the beet (*Beta vulgaris*), a species which includes the table beet for its edible roots or its greens in salads. For brewing, winemaking, or infusing we use the taproot, which contains high concentrations of sucrose. Swiss chard is also a type of *B. vulgaris*. If you are concerned about genetically engineered sugar beets, or the chemicals used to grow them, much like corn, growing your own is one way to ensure against GEO (mistakenly called GMO) lurking in the food you consume.

The beet can take some frost and prefers growing in the cooler temperatures of fall and spring. Be careful not to crowd them or the root will not fatten up into a globe.

Sugar Beets

BOTANICAL NAME: *Beta vulgaris*

PLANT TYPE: root vegetable

USDA ZONES: all zones. Where summers are hot (in parts of Zone 8 and warmer), grow beets as a fall, late winter, or early spring crop.

HEIGHT: 12 inches

SOIL: fertile loam. The soil should be amended, tilled, and all rocks removed from the area of the rows where the beets will be planted for unhindered root growth. At the bottom of each row, work in complete, organic fertilizer at the rate of 1 cup per 5 feet.

LIGHT: full sun

WATER: regular. Keep well-watered, especially when the fifth and sixth leaves show, which signals the growth of the root.

GROWTH HABIT: swollen root with edible leaves

PROPAGATE BY: seed sown in spring

SPACING: 4 inches. Sow several seeds in a cluster every 4 inches and thin to the strongest seedling when plants reach 2 inches.

MONTHS TO HARVEST: two

PRUNING: none

HARVEST: Dig sugar beets when they measure about 2 inches in diameter. Larger beets will become fibrous and lose their sweet flavor.

NOTES: Sugar beets can grow in a variety of soils and climates. Plant in rows 12 to 16 feet apart. Feed with fish emulsion every two weeks once the seedlings are 4 inches tall.

BEST USED IN: liqueurs, beer, wine

The beauty of growing beets for brewing or making liqueurs is you can use them fresh, or freeze or can your harvest and save for a later brewing/infusion date. Frozen beets will last up to one year in the freezer.

To freeze your beets, select ones that are no larger than 2 inches in diameter. Leave ½-inch stem on one end and a ½-inch taproot on the other. Leaving them on prevents the beet from bleeding during the cooking process. Cover the beets in water, bring the water to a boil, and cook twenty-five to thirty minutes until tender. Promptly cool the beets in cold water to stop the cooking process. Peel the cooled beets, and remove the stems and taproots. Cut into cubes and package in freezer bags or containers. Leave ½-inch headspace to allow for freeze expansion. Seal the container and freeze.

The amount of beets you use will vary and depend on the recipes you use the root for in beer, wine, or liqueurs.

Hot Peppers

Peppers—no matter if they are hot, sweet, or bland tasting—like growing in the heat. If the climate where you live doesn't give you hot summers, grow your pepper plants against a south-facing wall, or in a hoop house to increase the warmth around your plants. You can harvest your fruit at almost any stage of their growth. The heat of your peppers will be considerably greater the more mature they are at the time of picking. Once the pepper has turned its final color, pick it; waiting any longer causes the peppers to decline quickly. If you aren't sure when they are mature, the rule of thumb is to begin harvesting the first peppers 75 to 90 days after transplanting them. However, the hotter the climate, the sooner the peppers are ready. The easiest way to harvest most peppers is to use pruners to cut the fruit off the plant.

Generally, serrano peppers are picked while they are still green, and the cayennes and tabasco peppers are picked when they are red and a bit soft and come off the plant easily. Jalapeños should be their full size of around 3 inches (depending on the variety) and dark green when you begin harvesting

them. Later, when they display corking (tan streaking) they are good for pickling. Red is their final mature stage and that is when the flavor turns both hot and sweet. Habaneros' ripe color will vary according to variety, but are picked at their final stage color.

Same cautions apply when you are harvesting your hot fruit and want to rub your eyes. Don't! Wear gloves and keep those away from your eyes too.

Caution also applies to how many peppers to add to your batch of beer. Do you like just a touch of pepper heat and good flavor, or would you prefer more of a jalapeño twist? Maybe you prefer a breathe-through-your-ears sensation from the hottest, spiciest beer your mouth can handle. For your first venture into spicing up your brew, start with a recipe with good reviews. Some recipes call for adding peppers to the boil; others call for using them in the secondary, and some call for both. Adjust and experiment until you design your own recipe that matches your heat tolerance. If you find a batch is too hot, let the bottles age to mellow the heat. Still too hot? Use the beer in a chili recipe!

If you can't use your peppers right away, store them in a paper bag with their stems still attached, and place in the refrigerator where they will store up to a week.

Pepper plants

Peppers, Hot and Sweet

BOTANICAL NAME: *Capsicum annuum* (jalapeños, wax, cayenne, paprika, and bell peppers), *Capsicum chinense* (habanero)

PLANT TYPE: herbaceous perennials, grown as annuals in cooler zones

USDA ZONES: 4–13, hardy in Zone 11 and above

HEIGHT: 1–4 feet

SOIL: sandy loam amended with compost

LIGHT: full sun

WATER: regular

GROWTH HABIT: low, shrubby

PROPAGATE BY: seed

SPACING: 18–24 inches

MONTHS TO HARVEST: two to five from planting out

PRUNING: none

HARVEST: Most peppers reach peak flavor once they turn their mature color of red, orange, yellow, or purple.

NOTES: Peppers need at least 70 degrees Fahrenheit to grow and set flowers. However, if the temperature is higher than 90 degrees Fahrenheit or lower than 60 degrees Fahrenheit, the plants drop their blossoms. Peppers are only direct-seeded in the very warmest zones. Most gardeners buy starts or sow pepper seed indoors ten to twelve weeks before the last expected frost. Transplants are set out when nights are consistently at or above 55 degrees Fahrenheit, approximately two to three weeks after the last expected date of frost. Peppers make good container plants. In the ground or in containers, peppers benefit from a complete organic fertilizer worked into the soil before transplanting, which should be enough to see the plants through the season.

BEST USED IN: craft beers, liquers

Anaheim chili

CHAPTER 4

All Sorts of Wines

Growing Grapes and Other Fruits

Grape wine. It's a term as redundant as avocado guacamole or cabbage sauerkraut. No one asks the waiter to recommend a good "grape" wine or the store clerk to point the way to the "grape" wine aisle. Order the house white and you don't have to ask what fruit is filling your glass. When we talk wine, the role of grapes is implied. Even the names of our wines—chardonnay, merlot, cabernet sauvignon, pinot noir—come from grape varieties. Wine equals grapes, right? Well, hold on, Ernest and Julio. It depends on how you define wine.

If you're a member of the European Union, wine is a "product obtained exclusively from the total or partial alcoholic fermentation of fresh grapes." Grapes. Period. On the other hand, wine can be more broadly defined as a beverage made of any fermented plant matter: an interpretation that opens up a whole world of wine to home crafters. Good wine is ten to fourteen percent alcohol, crystal clear throughout its natural color, and retains a hint of the taste and aroma of the original fruit. But if it's not made of grapes, is it truly wine?

Yes, it is wine, but wine with a first name. If it's not made from grapes, it's "fruit wine" or "country wine" and bears the name of the fruit, or other plant, that begat it, such as raspberry wine or dan-delion wine. So why are grapes the wine default? It's because grapes are ethyl alcohol waiting to happen. The fruits are self-contained little packages of sugars, enzymes, and acids that are eager and able to transform into wine with little human intervention (although humans intervene plenty in the process). No other fruit is as spontaneously tipsy. Obviously, nature intended grapes for our enjoyment. Wine is their destiny.

Just because grapes are such a uniquely boozy fruit is no reason to let them hog all the glory. Well-made wine from ripe plums or tart rhubarb is a delight for the eye and palate. With a few additions and a bit of know-how, practically anything can be made into wine. Country wines are easily concocted by the home crafter with wine yeast, acids, yeast nutrient, equipment from brewer's shops, and sugar from the kitchen cupboard. Making wine from fruits and herbs you grow yourself is an excellent way to preserve not only the harvest, but the memory of spring blossoms and summer days. Homemade wine is a time capsule of a year in your garden, to be opened and enjoyed as a well-earned reward for a season past and in anticipation of the one to come.

WINEMAKING BASICS

Good wine is a harmony of acid, sugar, tannin, and all the nuances of flavor that occur during fermentation. It's a nearly miraculous conversion that—while not able to turn water into wine—does amazing things to fruit.

At first, the process of winemaking may seem daunting, but it becomes clearer when you realize the process comes down to a standard progression:

- Gather, prepare, and juice or crush the fruit or plant matter that will star in your wine.
- Amend the juice or must with water, sugar, acids, tannins, and a Campden tablet as needed. Add the started yeast.

- The primary fermentation begins. This is a short and very active fermentation that should last less than a week. The liquid is then racked into a carboy for . . .
- The secondary fermentation. This is a gradual process that can take up to a year.
- Bottling or re-racking. The wine is siphoned into sterilized bottles or a fresh carboy.
- Aging. The bottles or carboy(s) are placed in a cool, dark place and left to mature for several months.

Considering wine's divine nature, it's no surprise that winemaking calls for an immaculate inception: everything—from the fermentation bucket to the bottles—must be as sterile as possible. This zealous observation of sanitation is necessary to ward off bacteria and mold spores that will commandeer your fermentation for their own selfish interests, which are almost certainly not the same as yours. Purchase a food-grade disinfectant from a brewer's supply or use general purpose, household bleach mixed at a ratio of 1 tablespoon of bleach to 1 gallon of water. Apply the disinfectant or bleach solution liberally and extensively to all your winemaking paraphernalia. Commercial sanitizers require short soaks and little if any rinsing. A bleach solution requires a twenty-minute soak and a thorough rinsing as chlorine residue can leave your wine with taste-bud-shriveling nuances of tin foil.

Prepare the fruit, flowers, or herbs according to the recipe. The resulting "must," as this mash is known, or the liquid derived from it is then placed in a food-grade plastic fermentation bucket. Food-grade buckets can be purchased from homebrew suppliers. These buckets are white or clear—not colored—and not purchased from the local hardware store where there's no guarantee of food safety. Place your fermentation container on a sturdy stand two or three feet above the ground. This is for your convenience when the time comes to siphon off the contents.

A crushed Campden tablet can be added at this time. While not necessary, a Campden tablet clears the way to a finer product by eradicating those aforementioned microscopic agents of wine destruction. Once the resident yeasts have been killed, reliable cultured yeasts can be added. These wine yeasts can be introduced twenty-four hours after the Campden tablet; any sooner and the tablet's sulfur dioxide will inhibit the commercial yeast as well. Rather than simply dumping the yeast in dry, take the extra step of making a yeast "starter."

When the yeast goes in, it's time to add the other ingredients, such as sugar, water, and yeast nutrient. Yeast grows best in a twenty-two percent sugar solution. The sugar must be completely dissolved.

Giving Yeast a Head Start

To give your brew a boost and ensure consistent results, make a starter culture of your dry yeast. Twelve to seventy-two hours before your scheduled brew day rehydrate the yeast in warm water (95–105 degrees Fahrenheit) according to package directions. Proof the mixture with something sweet for the yeast to digest: Malt for beer, sterilized juice for wine or cider, honey for mead. Add a ¼ teaspoon of yeast nutrient and 2 teaspoons of sugar for every pint of mix. A pint of starter is sufficient for 5 gallons of wine or beer. If the yeast is healthy, the mixture should soon start to foam. Half a day to three days later, the starter can be incorporated with the other ingredients.

There's no need to be fancy: inexpensive white table sugar is commonly used (although superfine sugar dissolves more quickly). Water that has been boiled must be cooled to at least 80 degrees Fahrenheit to avoid killing the yeast. Avoid distilled, very hard, or heavily chlorinated water. A yeast nutrient is a good investment; it's a mix of B vitamins, amino acids, nitrogenous matter, and other agents that encourage robust yeast growth and discourage the fermentation from grinding to a halt (known by winemakers as becoming "stuck"). Now the primary fermentation will begin, and the container must be covered loosely with a lid or plastic wrap, kept at 70 to 75 degrees Fahrenheit for four days to a week, and stirred twice daily. A vintner's heating belt is useful for maintaining temperature. Unlike a heating pad, which can be damaged beneath a 60- or 70-pound fermenter, a heating belt wraps around the vessel.

The primary fermentation is complete when carbon dioxide bubbles stop rising. You should see a defined deposit at the bottom of the container (easier to see in a glass jug). At this point, transfer the liquid into one or more carboys. The choice of how many secondary containers to use comes down to your wish to fine-tune your wine in different ways, or how full each carboy can be filled. The secondary fermentation requires an absence of air, so the liquid needs to fill a container as fully as possible, ideally just below the bottom of the airlock. One carboy to hold the entire batch is typical, but two vessels work better if the combination leaves a minimum amount of headroom in each. Any remaining space must be topped up with water that has been boiled and cooled, or wine of the same type. On the other hand, if there's too much liquid, save the excess for future topping up.

The transfer is usually accomplished with the aid of a siphon or a spigot built into the fermentation bucket. Avoid splashing the wine as it will add excess oxygen. Pour the liquid through a sterilized strainer or straining funnel. If you're feeling particularly daring, you can strain through cheesecloth, but the resulting trouble and mess may convince you to spend a few dollars on a nylon strainer. Resist the urge to squeeze or press the juice from the fruit—especially high-pectin ones such as apples and blackcurrants—as it will encourage pectin haze. Once the carboy is filled and topped off, firmly insert a bung with airlock. Fill the airlock reservoir with sanitizer, water, or vodka. This is the beginning of the secondary fermentation, and the carboy should be kept at around 60 degrees Fahrenheit.

The secondary fermentation will burble along for two months to a year, depending on the type of wine. The surest way to judge the end of this stage is when all bubbling has ceased and the wine is clear. This is the time to take a hydrometer reading; the wine is ready when the reading is the same three days in a row. Some wines are racked and, subsequently, topped off one or more times during this second stage. They may also be "fined" with a product such as a silicic acid and gelatin combination to clear the wine. For a higher-quality libation, siphon into a fresh carboy (or oak barrel) when the specific gravity drops to .996 or below, and let your wine bulk age in a cool place for one month to a year according to the recipe. If the wine is still cloudy, as a last resort you can filter it with a brewer's filtration kit.

Now celebrate, for the time has come to bottle your wine. White wines can be stored in bottles of any hue, but the color of red wines is preserved best in green bottles. Before beginning bottling, a quick reminder: sanitize, sanitize, sanitize. Contaminated equipment is the number one cause of homebrewing failure. So let's clean those bottles. Bottles can be soaked in the same brewer's disinfectant as other equipment; however, they're also good candidates for heat treatment in the oven, which truly sterilizes all surfaces. Bottles must be completely clean before being placed in the oven. At 225 degrees Fahrenheit. the bottles should be microbe-free in two hours. Bring the bottles to temperature gradually, about 5 degrees Fahrenheit per minute, to avoid breakage. Don't preheat; just lay the bottles on their sides into a cold oven and turn the dial. If your bottles are thoroughly dry, you can bake them with their openings wrapped in aluminum foil. After they've cooled, they then can be stored in their sterile state until called upon.

You can load your wine directly into the bottles via a siphon or spigot and tubing, or the liquid may first be drained into a pitcher (by one of these methods) and then poured into each bottle with the aid of a funnel. In the bottle, the wine should reach to about ¾ inch below where the bottom of the cork will sit. Use new corks and set them so that

Talk Winemaking like a Pro

Aperitif: a wine to be served before dinner; usually high in alcohol and slightly sweet

Brix: the percentage of sugar in the must

Cane: one-year-old grape growth

Cap: the head of foam and residue that forms atop a liquid undergoing primary fermentation

Chaptalization: neutralizing acid in the must or adding sugar to raise the alcohol level

Clarifying: the removal of "wine haze" caused by excess pectin

Country wine (fruit wine): wine not made from grapes

Cordon: a branch that grows horizontally from the top (or, in some systems, top and center) of the trunk of a grapevine. Fruit-bearing shoots grow from the cordons. Most trellis systems have two cordons, extending in opposite directions along a horizontal wire.

Dessert wine: sweet wines to accompany the final course of a meal. Paired with fruit or dessert.

Drupe: a stone fruit, like a plum or cherry

Filtering: running the wine through a commercially available filtration kit to remove or reduce wine haze

Finings: a clearing agent that clarifies wine or beer near the end of the secondary fermentation; to "fine" is to clarify the solution

their tops are level with the top of the bottles. Leave the bottles to sit upright for two days, and then put them on their sides to keep the corks from drying and shrinking. Store as you would store-bought wine—in a cool, dark, dry room or cellar.

Once in the bottle—or, alternatively, stored in a carboy sealed with a plastic bung—your wine will embark on its final maturation, a "reductive aging." This stage gradually uses up the oxygen within the bottle. Although the environment is nearly

Haze: cloudiness in wine caused by suspended particles such as pectin or protein

Macerate: to crush and steep fruit without heating

Must: a fruit or vegetable mash or solution containing sugar and headed for fermentation

Oenology: the science of winemaking

Pectinase: an enzyme that breaks down pectin to clarify wine

Pétillant: sparkling characteristic of wine

Reductive aging: maturing that occurs after wine is bottled. It is "reductive" because the amount of oxygen in the bottle is gradually depleted.

Straining: running the wine through a sieve or cheesecloth to remove impurities. For an extra-clear wine, line the sieve with cheesecloth.

Table wine: wine to be served with dinner

Véraison: the onset of ripening when the grapes begin to change color

Vigneron: a grape grower

Viniculture: the study and production of grapes specifically for wine

Vinification: winemaking

Vintner: a winemaker

Viticulture: the study and production of grapes

airtight (a minute amount of oxygen is absorbed through the cork), enough oxygen is present to react with compounds in the wine for many years, allowing some wines to develop complex flavors and aromas. Study your recipes carefully, though; while extended storage is a plus for many wines—including many grape ones—certain country wines are an exception. Most wines do benefit from several months of maturation in the bottle. However, if you fail to drink your country wine within a year, it may begin to lose quality.

GRAPES

Welcome to the Grand Order of the Grape. Thinking of becoming a member? As clubs go, it's not too demanding; there's no secret handshake, but we do have our own arcane vocabulary: Cordon, vigneron, phylloxera, véraison. Our mascot? The four-armed Kniffen. And the initiation! Not to give too much away, but you're going to need a sturdy pair of pruners. We even have a club creed: spur pruning keeps cordons year to year, but vine pruning cuts the cordons near (the trunk).

Let's check the handbook to see if you'd like to join.

Grapes like it hot, dry, and sunny. Although these conditions are optimal, one type of grape or other can be grown in almost any area of the United States. In colder climates, give grapes the advantage of every bit of sun and heat you can wring out of your site, such as a south-facing situation with a wall at their back or a frost-pocket-free slope. The choice of variety is extremely important, not only for the type of wine you'd like and the amount of space you have, but ultimately for vines that will survive and thrive in your backyard—wherever that may be.

Four types of grapes can be used for wine: European (*Vitis viniferia*), northeastern American (*V. labrusca*), southeastern American muscadine (*V. rotundifolia*), and

A grape arbor can be a defining feature of a patio.

interspecific hybrids between European and American grapes (known as French-American hybrids). Northeastern American grapes (such as Concord) are the most cold-hardy, while European grapes, which make up the vast majority of commercially produced wine, do best in regions with Mediterranean-type climates. Although there's no question that European grapes are the gold standard for wine production, they are also the most finicky.

Muscadines are well-adapted to the heat of the south. They make non-standard, regionally enjoyed wine that's considered by connoisseurs to be a bit "assertive." The French-American hybrids are generally hardier than European varieties, if not quite so much as American species. They're also nearly as disease-resistant as their American kin, while the wine they produce compares with that of their European side. Within each of these broad categories of grape are multiple cultivars. With so many available varieties, the aspiring vintner with an optimal climate faces an exciting—if head-spinning—choice. Do your research and ask local experts what grapes perform best in your region.

Buy certified, virus-free, one-year-old plants. Year-old plants are less likely to be last year's rejects and more likely to survive transplanting than older plants. When your plants arrive, do not allow the roots to dry out. Before planting, which should be within a week of delivery, soak the roots in water for six to eight hours. As you sur-

Grapevine climbing up a tree

vey your property, deciding where to plant your grapes, keep in mind not only sun and heat, but longevity. Grapes are long-lived and need a permanent location with plenty of room for an 8-foot run per plant (or 16 feet for muscadines). Grapes appreciate well-drained, slightly alkaline soil. The planting area shouldn't be too nutrient-rich, or the vines will put their efforts into leaves at the expense of fruit.

Of course, your vineyard will need a sturdy and permanent support. Single vertical poles, like those used in growing hops, are inappropriate for grapes, which need a long run. Although you can grow grapevines on a fence, arbor, or pergola, the most convenient and businesslike support is a trellising system of poles and wire. Your most important decisions when planning your trellis are how high, and one tier or two.

Two-tier (two-wire) trellises are good, all-purpose systems for grapes. They are made with sturdy 8-foot wood or metal

Grape

BOTANICAL NAME: *Vitis vinifera*; *Vitis labrusca*

PLANT TYPE: woody vine

USDA ZONES: 4–10

HEIGHT: Vines can grow to 115 feet in length, but seventy to ninety percent is pruned away. Trunks are formed at 1–6 feet, depending on the pruning system.

SOIL: deep, well-drained, slightly acid. Soil that is too rich encourages plant growth at the expense of fruiting.

LIGHT: full sun, hot site

WATER: medium to dry. Grapes tolerate drought once established.

GROWTH HABIT: spreading vine, climbing by coiling tendrils

PROPAGATE BY: cuttings, grafting

SPACING: 5–8 feet

YEARS TO BEARING: three—remove all flowers for the first two years.

TRAINING: Specific to grapes, new grape plants should go in the ground in spring as short sticks with only three to four buds. Do no further pruning the first summer as the buds grow into shoots. That winter, cut away all but the sturdiest shoot; trim it to four buds, and tie it to the upright support. This is the start of the trunk. The following spring, allow offshoots from the new trunk to grow to no more than 8 inches before selecting the strongest to continue the trunk; tie this chosen shoot upright to the support. Remove the other shoots, unless you have a two-tiered (four-arm Kniffen) support, then also keep two strong lower shoots to head along the bottom wires (one to each side). Cut off all other shoots. Top the trunk shoot (cane) at 8–10 inches above the top wire—wherever you chose the top wire to be. The height of the trunk can be 1 foot (low wire) or 4–6 feet (high wire) depending on your preference, and whether your chosen grape variety tends to hang down or grow upright. Now the vines growing from the trunk will be pruned according to the system (cane or spur) and the type of support (wire or arbor).

PRUNING: Grape fruit is produced on new growth. Grape pruning is done annually during the dormant

posts set 24 feet apart and 2 feet into the ground. Two wires are strung along the posts: the lower wire 40 to 45 inches above the ground, the top wire 12 to 14 inches above the lower wire. End posts carry most of the stress, so these need to be especially sturdy and set at an angle against the pull of the wires.

A two-wire system such as this is made for four cordons (branches): two season (before the buds swell) for the purpose of limiting the amount of fruiting wood, thus lessening shade and limiting the load of fruit, which maximizes quality. There are different pruning systems, which are appropriate for different grape cultivars. All pruning systems are done atop the trunk developed during the first two years' training. The cane system of pruning works well for most grape types and most at-home growers. During dormancy, cut off all but two canes (or four, on two-tier Kniffen systems) of last year's growth, including last year's two (or four) horizontal canes, known as cordons. The chosen canes will come from the old cordons; they should be strong new shoots growing as near as possible to the trunk. These will replace last year's cordons (which you just cut away) and are bent over and wound around the wires (or through the arbor, if that is your chosen support). They are trimmed to ten buds. From these new cordons will grow new, fruit-producing canes. In this form of pruning, the old cordons are cut away each year and replaced by two new canes (or four, in the Kniffen system) to become that year's cordons; only the trunk remains the same from year to year.

THINNING AND HARVEST: To thin the fruit, space the clusters so they don't touch. Harvest is approximately 5–30 pounds in September and October. Grapes may change color before they are fully ripe, so a taste test is the best indicator of ripeness. Cut the clusters when they are dry, and keep them that way until it's time to put them in the fermenter. Grapes keep their quality for two to three weeks in the refrigerator.

NOTES: Well-tended vines can produce thirty to sixty clusters. Grape plants are self-pollinating. They live eighty years or more. Remove flowers for the first two years to allow the plants to develop strong root systems. Many animals love grapes.

BEST USED IN: wine

Fresh grapes on the vine

putting the plants in the ground to avoid damaging them.

Grapes also can be grown on a decorative trellis, arbor, or pergola. A trellis is a two-dimensional support; it can be free-standing or attached to a building. An arbor has a roof and "walls." Arbors often arch over a walkway or incorporate a gate or bench. A pergola has posts (four, at least) holding up a roof. Pergolas may provide shade or serve as outdoor extensions to a home's sitting or eating area. Be aware that grapes grown on these ornamental structures can be lovely, but they're also challenging to prune, especially if latticework is involved. If you're planning such a support, it's prudent to leave out the lattice and stick to clean, uncomplicated posts, beams, and rafters. Even grapevines grown for ornamentation and shade must be pruned.

Once the trellis is up, it's time to plant your vineyard. Amend the planting area with compost and, as long as you're digging around, remove every trace of weeds. Grapes are usually planted in the spring, before they break dormancy. Plant 8 to 10 feet apart. Dig the holes to such a depth that the graft union sits 3 to 4 inches above the ground. While the roots are visible, install a 1x1-inch wooden (or bamboo) grape stake at the base of each plant; stakes should be about 5 feet long with 1 foot of that length in the ground for a sturdy hold. These will be used to stabilize the canes you select to become the trunk.

per wire, one to each side. A single-wire system is another common choice; in this case, the wire is set anywhere from 1½ to 6 feet above the ground. The height of the top wire will be the height at which you develop the trunk, which is determined in large part by the growth habit of the grapes you choose. Use 9- to 12.5-gauge fencing wire and posts of metal or rot-resistant wood. Build your trellis before

Cover the roots with soil and fill the hole with water to the top. As the water soaks in, it will settle the soil firmly around the roots. Finish covering with soil, leaving a depression to hold water and make the job easier. If the plants didn't arrive pre-pruned, prune them above the graft to two to three buds on the strongest shoot and remove all other shoots and buds. Also, remove any stray roots coming from above the graft. Keep the soil moist, but not soggy, for the first month.

Soon the shoots will begin to live up to their name. As they rocket upward, keep the new canes tied to the stake; vines flopping around on the ground are an invitation to disease and damage. If any precocious flower clusters appear the first or second year, nip them in the bud. The job now is to train the vines to a trunk and—if you'll

A Clear Improvement: How the Romans Made Wine Better through Glass

"The first taste is in the eye."

Sophocles—a Greek—may have said it, but the Romans perfected it, at least when it comes to wine. And it came about through glass.

Although Romans did not invent glassblowing, they greatly improved it. Not only did the Romans develop innovative new glassblowing techniques, they also discovered that clear glass could be achieved at much lower temperatures with the inclusion of flux—an agent added to reduce the melting point of the quartz then in use.

From this breakthrough, delicate, thin-walled drinking glasses were born. Now the citizens of Rome—used to imbibing from opaque ceramic, metal, or horn vessels—could see what they were drinking. Suddenly, the color and clarity of their wine became a consideration. Ever since, the world has sipped with both tongue and eye.

be spur pruning—permanent cordons. The first winter after planting, cut away all but the sturdiest cane; trim it to four buds and tie to the upright support. This is the start of the trunk. The following spring, allow offshoots from the new trunk to grow to no more than 8 inches before selecting the strongest to continue the trunk; tie this chosen cane upright to the support. Remove the other canes, unless you have a two-tiered (four-arm Kniffen) support; then also keep two strong lower canes to head along the bottom wire; one to each side. Cut off all other canes. Top the trunk cane at the top wire—wherever you've chosen the top wire to be. The height of the trunk can be 1 foot (low wire) or 4 to 6 feet (high wire) depending on your preference and whether your chosen grape variety tends to hang down or grow upright. Now the vines growing from the trunk will be pruned according to the system (cane or spur) and the type of support (wire or arbor).

Grapes are produced on new growth. Pruning is done annually during the dormant season (before the buds swell) for the purpose of limiting the amount of fruiting wood, thus lessening shade and limiting the load of fruit, which maximizes quality. There are different pruning systems, which are appropriate for different grape cultivars. All pruning systems are done atop the trunk developed during the first two years' training. The main structural pruning is done in the dormant season shortly before growth starts.

Cane pruning and spur pruning are the two main forms of grape training. Research any grape variety you plan to grow to learn which pruning style it demands. If you have mystery grapes, start with cane pruning; if the fruit appears close to the cordons, from the first couple of buds of last year's growth, switch to spur pruning. If clusters of grapes are spread out all along the cane, continue to cane prune.

The cane system works well for most grape types and most at-home growers. To cane prune, cut off all but two canes (or four, on two-tier Kniffen systems) of last year's growth, including last year's two (or four) horizontal canes, known as cordons. The chosen canes will come from the old cordons; they should be strong new shoots growing as near as possible to the trunk. These will replace last year's cordons (which you have just cut away) and are bent over and wound around or tied to the wires (or through the arbor, if that is your chosen support). They're trimmed to ten buds, leaving approximately 18 to 24 inches of vine. From these new cordons will grow new, fruit-producing canes. In this form of pruning, the old cordons are cut away each year and replaced by two new canes (or four, in the Kniffen system) to become that year's cordons; only the trunk remains the same from year to year.

Americans in Paris, or, Gee, Thanks

Not everything should be shared. In the mid-1800s, European vineyards became a stunning example of the worst that can happen when non-native species are introduced to a new continent.

From the earliest days of farming in the New World, grapevines imported from Europe withered and died in the foreign soil. No one was sure of the cause, but whatever was making the European vines fail didn't appear to affect America's native grapes (which made disappointing wine). Unfortunately, botanists ignored the wine-red warning signs and eventually sent samples of the robust native grapes to Europe.

Thus were the gates opened and the Trojan horse rolled in. American soil, it turns out, is infested with nearly microscopic sucking insects that feed on grape roots, causing them to deform, choking off water and nutrition, and opening the door to other organisms that join in the assault. Eventually the tiny parasites were discovered and given the name phylloxera (*Daktulosphaira vitifoliae*). With the native soil infested by these greedy little organisms, American grapevines had to adapt or die, and adapt they had. Although the native vines are attacked by phylloxera, the plants have evolved to flourish in spite of the pests.

Not so with European grapes. When the American vines were introduced to the Old World, the devastation to French vineyards was swift and very nearly complete. Within twenty-five years, three-quarters of European vineyards had succumbed to phylloxera. As the vines died, viniculturists scrambled to save the wine industry. The solution came from grafting: European cultivars were affixed to phylloxera-tolerant roots. More recently, hybrids of European and American grapes are showing promise in both vigor and vinification.

Spur pruning is the other basic style of grape training. It differs from cane pruning in that the cordons are permanent rather than being replaced each year. Knobby spurs are developed every 4 to 6 inches along the cordons; from these grow the fruit-producing canes—usually two canes each annually. Each year late in the dormant season, prune to one single cane per spur. That cane is then cut back to the first visible bud from the base. In the spring, that bud will grow into a cane with two clusters of grapes; a second bud, hidden at the base of the cane stub, will grow to be next year's keeper cane. The keeper cane should produce a single cluster of grapes.

To the uninitiated, grape culture might seem an elite and elaborate business. With a bit of practice, though, caring for your vineyard soon will be a seasonal routine.

Epic Error

Let's raise a glass of champagne and make a toast to—well—mistakes. May all of our mistakes be as brilliant as this.

From the Middle Ages until the 1800s, sparkling wine was an embarrassment to self-respecting vintners. No one wanted a beverage befouled by bubbles. Fortunately, try as they might, cellar masters were unable to rid the world of the effervescent menace. The cause of the wayward wine may have been one thing vintners couldn't control: the climate. Beginning at the end of the 1300s, increasingly cold winter weather hindered the ability of yeast to consume all of the sugar during primary fermentation. By spring, the wine had been bottled. But it wasn't finished. Fermentation resumed with the rising temperature. Now there was no escape for the carbon dioxide naturally released when sugar converts to alcohol. Voilà! Bubbly!

Eventually the Champagne region of France convinced the world that sparkle was special and an industry was born. So here's to serendipity. Sometimes good fortune is really a mistake made by those with vision (or a good press agent).

FRUIT WINES

What type of country wine pops your cork? How about a blackberry Bordeaux? Or a blueberry burgundy? Maybe an upcoming special occasion calls for homemade peach champagne? One thing is certain; if you make wine from your homegrown fruit you will not only be the envy of your friends, you'll be surprised at how many friends you have.

Wines made with fruit, herbs, and flowers have long been staples in cooler regions where grapes grow grudgingly, if at all. In tropical climes, banana wine is common. Practically anything that can be fermented can be made into wine. With any plant matter other than grapes, you will be called upon to add or adjust the yeast, sugars, acids, and tannins. As an example, strawberries and raspberries are high in acid when turned into wine, and their must (fruit pulp) needs to be cut with water before fermentation, then "back-sweetened" after fermentation to avoid leaving the imbiber with a permanent pucker. On the other hand, blander fruit such as blueberries and dessert pears (but not perry pears) need an acid boost.

Are there any limits on what can be fermented? Technically, if you can grow it at home, you can make it into wine. But, honestly, can you imagine asking your guests to choose between a grass-clipping sangria and an onion cordial? That's not to say your creativity can't run a bit wild. Blending lets you put your personal stamp on truly tasty libations. A good way to concoct your combinations is by considering those that work in the kitchen—like apple and cranberry, blackberry and lavender, rosemary and black currant. However, until you become a winemaking virtuoso, you'll want to rely on a proven recipe.

It's important to know that no two batches of wine, even from the same recipe, will turn out the same. There are many considerations that go into planning your homegrown, homebrewed wine: Red wines are best made from must, while white wines often begin with only the rendered juice. Country wines are challenging compared to grape wines because most fruits contain a plethora of pectins. This gelatinous substance, used to set jam, will cloud your wine and needs taming with a pectin enzyme. Cooking or boiling fruit prior to fermentation can set the pectin and increase the tendency of your wine to haze.

Many fruit wine recipes call for the addition of grape juice or grape concentrate to round out the final product. Clear fruit juices or frozen grape concentrate purchased from the grocery are fine, but leave the red juice on the shelf as it may be artificially colored, which won't translate well to your post-fermentation product. Finally, flowers and herbs alone tend to result in delicately flavored wines. When used as an adjunct to fruit wine, rather than including them in the primary fermentation—where their essence is more easily lost—consider adding them during the secondary fermentation once the specific gravity reads 1.000.

Extracting Flavor

You might recognize this as the same technique as dry-hopping beer. The flavor of some flowers and herbs can be coaxed out by boiling in water, seeping, and straining; once cooled, the liquid is added to the secondary fermentation. Certain fruits also can be introduced later to impart more of their fresh flavors to the wine.

There are several methods of extracting flavor from fruit, flowers, and herbs. The simplest way is to add the plant material to the bucket prior to primary fermentation. Alternatively, it can be steeped in boiling water on the stove or mixed with boiled water directly inside the fermentation bucket before anything else is added. Finally, it can undergo cold maceration in which the herbs or flowers are covered with cool water and left to soak overnight; add a dose of pectin enzyme and the maceration is complete two days later. Most fruit goes into the fermentation bucket uncooked—but a few, including microbe-friendly elderberry—may be brought to a boil and simmered for several minutes before being cooled and fermented. Don't steam fruit, as that will cause it to oxidize.

Of course, before you can start extracting their essence, you must collect and clean your fruit, flowers, and herbs. Fruit should be perfectly ripe, clean, and sound; if you wouldn't put it in your mouth, don't put it in your fermenter. Windfalls are okay, if you meet the preceding criteria; wash them to remove any contaminants, and pare away bruises. It doesn't hurt to wash all firm fruits, especially if there's any chance of pesticide residue. Berries such as raspberries and blackberries are delicate and resent manhandling, but if there's any reason for concern, soak them in water and drain. Herbs can be swished in a bowl of water. Keep in mind that with whatever you're gathering, you should harvest a bit extra of it, just in case.

After cleaning, consider freezing your fruit. Even an overnight freeze will burst cell walls and release additional juice. This is an especially useful tactic for berries. Once they're thawed, the berries can be easily crushed with a potato masher. Avoid using a blender, which will both oxidize the fruit and pulverize seeds, releasing a bitter taste. Soft fruits such as peaches, plums, and apricots don't require crushing; simply cut the fruit into chunks, remove the pits (or leave them in, unbroken), and into the fermenter they go.

When planning your wine garden, decide what kinds of wine you prefer. After all, your space and time are likely limited. Are you a fan of berry wines (which can cost a fortune commercially)? Maybe a blackberry or blueberry patch is for you. Do you like white wine but have room for only one tree? Consider an apricot, if you have the climate. Of course, climate should be your first consideration when choosing any plant. Even within a species, different cultivars are usually better adapted to different conditions. When in doubt, ask an expert in your area for recommendations.

Apricot

BOTANICAL NAME: *Prunus armeniaca*

PLANT TYPE: deciduous small trees. Available as dwarfs, semi-dwarfs, and standards

USDA ZONES: 5–8

HEIGHT: dwarf, 8–10 feet; standard, 15–20 feet

SOIL: well-drained, light and loamy soil. Apricots can tolerate somewhat alkaline soil.

LIGHT: full sun, part shade

WATER: regular. Apricots tolerate drought, but may drop fruit if water is inadequate during fruit development.

GROWTH HABIT: vigorous; needs heavy pruning

PROPAGATE BY: grafting, seed

YEARS TO BEARING: dwarf, two to three; standard, three to five

THINNING AND HARVEST: Thin clusters to 6 inches apart, one fruit per cluster, in mid-spring when the embryonic fruit is no bigger than a nickel. Apricot trees bear 1 to 5 bushels. Store apricots on the kitchen counter for four to five days or in the refrigerator in a plastic bag for a week. Don't wash until you're ready to cut them for the fermenter.

NOTES: Most apricots are self-fertile, but some need a pollinizer. Use little or no fertilizer to discourage overly vigorous growth. Prune lightly for the first five years to encourage fruiting; switch to heavier pruning once trees begin to bear. Fruiting spurs bear for about four years, at which time they should be removed. Apricots can be espaliered. They are susceptible to frost injury and require a long growing season, but they also need a chilling period of 500–900 hours below 45 degrees Fahrenheit. The trees are very early blooming, and the buds may be damaged by spring frost. Apricots are more drought tolerant than many fruits. For 5 gallons of wine use about 18 pounds of apricots.

BEST USED IN: delicate table wines and champagne. Favored by some as a second choice to grapes as a quality white wine. Apricots tend to harbor natural yeast nutrients, so ferment readily. When processing, the pits may be left in to go into the fermenter as long as they aren't cracked or crushed, which can cause bitterness in the wine.

Black currant

BOTANICAL NAME: *Ribes nigrum*

PLANT TYPE: upright shrub

USDA ZONES: 3–8, best in Zones 5 and colder

HEIGHT: 4–5 feet

SOIL: average, loamy garden soil. Will tolerate heavy or sandy soils

LIGHT: full sun or part shade in warmer areas

WATER: regular

GROWTH HABIT: multiple-stemmed, thornless

PROPAGATE BY: cuttings

SPACING: 3–5 feet

YEARS TO BEARING: two (three to four years for full production)

PRUNING: When plants are four years old, prune before growth starts to remove oldest branches. Each year after that, remove canes that are more than three years old during dormant season. Fruit is produced on new wood.

HARVEST: 3 to 10 pounds per plant. Pick when fully ripe for best flavor. Black currants ripen individually within clusters, so must be picked individually. Avoid washing currants, as they are susceptible to fungus. Can be kept in the refrigerator for two weeks.

NOTES: Currants are illegal to grow in some areas because they host blister rust, which kills American white pines; check restriction in your area and buy only rust resistant varieties. Many black currant varieties require a pollinizer; use another black currant variety. Red and white currants do not require a pollinizer. All currants are very cold hardy and will not perform well in warm-winter climates. They resent heat. The plants are productive for about fifteen to thirty years. For 5 gallons of wine use about 12 pounds of currants.

BEST USED IN: red wine and sweet dessert wines, liqueurs, and cordials. The strong taste of black currant is a good addition to ciders.

Blackberry

BOTANICAL NAME: *Rubus fruticosus*

PLANT TYPE: biennial bramble

USDA ZONES: 5–9

HEIGHT: 10–12 feet without pruning. Canes should be trimmed to 4–6 feet

SOIL: deep, well-drained. Mulch is beneficial.

LIGHT: full sun

WATER: regular watering through the growing season

GROWTH HABIT: upright or trailing vine. Perennial roots bear biennial canes. Some cultivars are thornless.

SPACING: rows 10 feet apart with upright varieties 2 to 3 feet between plants, and trailing varieties 5 feet

PROPAGATE BY: canes that sprout from the roots. Buy certified disease-free plants. Plants from a friend's garden may carry disease.

YEARS TO BEARING: two

HARVEST: 1 to 4 quarts per plant. Blackberries are ready to pick when the fruit slides easily from the plant.

NOTES: Blackberry canes should be trained onto a trellis. Prune 6 inches off the tips of upright blackberries in summer, once the canes are 3 to 4 feet; cut the resulting side branches to 1 foot. In fall, remove canes that have fruited. Also remove diseased or weak canes and those growing outside the designated bed. Thin the remaining canes to 12 inches apart. Unlike black raspberries, blackberries retain their core when picked. The Himalayan blackberry *Rubus armeniacus* is invasive in some areas. For 5 gallons of wine use about 15 pounds of blackberries.

BEST USED IN: Possibly the finest of the berry wines, blackberries make a full-bodied red wine, which, if aged correctly, resembles a Bordeaux.

Blueberry

BOTANICAL NAME: *Vaccinium* species

PLANT TYPE: deciduous bush. A few varieties are evergreen.

USDA ZONES: 4–8. Some lowbush varieties are hardy to Zone 3; a few others are specially bred for Zone 10.

HEIGHT: 1–15 feet

SOIL: Rich, moist, acidic (pH of 4 to 5.5), high in organic matter (especially peat)

LIGHT: full sun

WATER: regular to liberal. Will need to be well-watered the first summer and given supplemental water in subsequent summers.

GROWTH HABIT: shrubby

PROPAGATE BY: softwood cuttings (difficult to root)

SPACING: same as the height of the plant. Lowbush, 1½ feet apart in rows 3 feet apart; highbush and rabbiteye, 6 feet apart in rows 7–9 feet apart; hardy half-high hybrids, 5 feet apart in rows 6–8 feet apart.

YEARS TO BEARING: four to seven years for full production

PRUNING: Once the plant is five years old, cut back older branches to new growth. Occasionally remove the oldest branch to the ground.

Blueberries' fall colors are stunning.

Blueberry (Darrow) flowers

HARVEST: 2–25 pounds depending on type and variety. The berries hold well on the plant, so can be picked in large lots once a week (if the birds don't get them first). Pick when fully ripe for best flavor.

NOTES: Blueberry plants are attractive in the landscape. Different types of blueberries are best adapted to different regions. The long-lived plants can produce for eighty years. Grow at least two cultivars for best pollination. Blueberries need cool winters. Feed blueberries in spring just before growth begins with ammonium sulfate or a complete, acid-based fertilizer in a formulation as close to 16-8-8 as possible. Follow package directions and don't over-fertilize. An annual topdressing of an acidic mulch such as peat moss, cottonseed meal, or conifer needles is beneficial. Birds love blueberries too; netting is needed. Some blueberries have high potassium sorbate levels that inhibit fermentation. For 5 gallons of wine use about 13 pounds of blueberries.

BEST USED IN: mild, sweet red wine

cherry blossoms

Cherry

BOTANICAL NAME: *Prunus avium* (sweet cherry); *Prunus cerasus* (sour or pie cherry)

PLANT TYPE: tree

USDA ZONES: 5–9 (sweet); 4–9 (sour)

HEIGHT: standard sweet cherries, 20–40 feet; sweet cherries on dwarfing rootstock, 10–12 feet. Sour, 15–20 feet

SOIL: light and well-drained with mid-fertility

LIGHT: full sun

WATER: regular. Cherries cannot stand in waterlogged soil, but neither should the roots dry out. Water during arid summers.

GROWTH HABIT: upright tree (sweet cherry); spreading tree (sour cherry)

PROPAGATE BY: grafting, seed

SPACING: for sour and dwarf sweet cherries, 12 feet; for standard-sized sweet cherries, 20–25 feet.

YEARS TO BEARING: sweet, five to seven; sour, two to three

PRUNING: Prune after flowering to reduce chance of bacterial infection. Cut back mature sweet cherries only to thin, or remove broken or diseased branches. Completely remove the occasional older branch of a mature sour cherry to encourage new growth. The fruit grows on spurs that are long-lived don't need to be renewed by pruning.

HARVEST: 25–50 pounds. Sour cherries ripen two to three weeks after sweet cherries. Pick when the fruit is fully ripe; cherries don't ripen further after harvest. Cherries can be frozen over several weeks as they are gathered. Freezing releases more juice from the cherries for fermenting.

NOTES: Cherries are best in regions with chilly winters and mild to moderate summers. Cold, wet springs can greatly reduce the crop. Some sweet cherries are self-fertile. When choosing two sweet cherry cultivars, be certain they will cross with each other. All sour cherries self-pollinating, but won't pollinate sweet cherries. The trees have a life expectancy of thirty-five years. Birds love cherries. A quart of pitted cherries makes a gallon of wine.

BEST USED IN: sweet red wines. Cherries ferment well. Occasionally, a variety of sweet cherry can produce a slight off-flavor in wines; Bing cherries don't have this fault. Neither do sour cherries, which are considered to make the finest of the cherry wines. The pits don't need to be removed prior to fermenting; however, do not crush the pits. Remove the stems.

Cranberry

BOTANICAL NAME: *Vaccinium macrocarpon*

PLANT TYPE: evergreen ground cover

USDA ZONES: 3–9. If temperature goes below 10 degrees Fahrenheit, plants need to be heavily mulched.

HEIGHT: 2–6 inches

SOIL: well-drained, very acidic soil, high in organic matter. Add peat or sand for good drainage. Cranberries can be grown in straight peat moss. Consider removing existing soil and replacing with a customized mix especially for cranberries. No bog needed.

LIGHT: full sun

WATER: heavy. Soil must remain moist.

GROWTH HABIT: spreading stems root to increase the patch

PROPAGATE BY: self-layered stems

SPACING: 1 foot apart in rows 2 feet apart

YEARS TO BEARING: three

PRUNING: Cranberries are "pruned" with sand. Every two to three years, ½ inch of sand is used to cover the ground-lying runners, leaving the upright growth standing above. This is done just at the start of the growing season. The runners can be cut back to keep an orderly bed, but never prune the fruit-producing uprights.

HARVEST: about 1 pound per every 5 square feet within seven years. Cranberries are harmed by winter freezing, so pick the fruit before the first frost. If they must be harvested before they are fully ripe, the berries can be laid out in a warm spot indoors to finish ripening. The berries are ready for picking when they're dark red with brown seed. The berries must be hand-harvested. The fruit can last two months in the refrigerator.

NOTES: Because the plants are frequently replaced by rooting runners, cranberry beds can survive and produce almost indefinitely. The beds require careful weeding because of their low stature. The heavy peat moss component of the cranberry bed will discourage many—but not all—weeds. Cranberries aren't grown under boggy conditions; they are sometimes commercially harvested by flooding (creating a temporary bog). Feed first-year cranberry plants with a high-nitrogen fertilizer to encourage runners. Discontinue the nitrogen by years two to three to let the berry-producing uprights to take over. In cold winter areas, cover the bed with a heavy mulch or row cover. Wine recipes may call for 2–3 pounds of cranberries.

BEST USED IN: red wine, especially sweet wine. Also a good addition to cider.

Dandelion

BOTANICAL NAME: *Taraxacum officinale*

PLANT TYPE: tap-rooted, herbaceous perennial

USDA ZONES: pretty much everywhere between the arctic and the tropics.

HEIGHT OF FLOWERS: 2–18 inches

SOIL: any

LIGHT: full sun to light shade

WATER: medium to dry

GROWTH HABIT: ground-hugging basal rosette

PROPAGATE BY: seed-spacing, 1 foot apart to all sides

MONTHS TO BEARING: 2–4

PRUNING: Pull off seedheads (dandelion "clocks") to prevent invasive reseeding.

HARVEST: Collect dandelion flowers on a sunny spring day (March through May) when the flowers are fully open. With a pair of scissors, trim the yellow blossoms from the green calyx and discard the calyx, which adds bitterness to the wine. Plan to pick enough flowers to loosely fill a gallon container, but check your recipe and remember that you'll need extra to make up for the discarded calyx.

NOTES: It's not news that dandelions self-pollinate. The flowers are usually boiled or covered in boiling water and left to steep for two to four days. The resulting infusion is then fermented. A quart of dandelion flowers makes about a gallon of wine.

BEST USED IN: a standard, easy, country white wine: rich and medium sweet. All parts of the plant are edible, but only the flowers are used in fermenting.

Elderberry/Elderflowers

BOTANICAL NAME: *Sambucus* species;
most common for fruit, *S. nigra*

PLANT TYPE: shrub

USDA ZONES: 3–9

HEIGHT: 6–15 feet. Taller varieties can
be pruned to 8 feet.

SOIL: slightly moist. Elderberries resent
dry soil

LIGHT: full sun or part shade in
warmer areas

WATER: regular. Must have water during
driest part of summer.

GROWTH HABIT: sprawling bushes

PROPAGATE BY: seed and root suckers

SPACING: 6–8 feet

YEARS TO BEARING: two (three to four
years for full production)

PRUNING: Prune before growth starts
when plants are four years old to
remove oldest branches.

HARVEST: 8–10 pounds of fruit per plant.
Pick only ripe elderberries; unripe
berries are mildly toxic. Elderflowers
should be picked when creamy white
with flowers that are open, but haven't
begun to dry. Pick flowers at the end
of a warm day for maximum flavor.
Elderflowers are too delicate to be
washed, so simply shake out insects
and place flowers outside in a shady
spot for up to an hour to allow any
remaining insects to escape.

NOTES: Elderberries produce better with
cross-pollination; plant two *S. nigra*
cultivars. The bushes are very cold
hardy. Don't use the fruit of wild red
elderberries, which are toxic. Birds
love elderberries too. Elderberry
bushes are attractive in the landscape. For 5 gallons of wine use
about 10 pounds of elderberries.

BEST USED IN: a commonly produced,
fine red wine. The flowers also make
a delicate white wine, sparkling or
not. The berry juice is a good addition to cider and soft drinks.

Gooseberry

BOTANICAL NAME: *Ribes uva-crispa* (European); *Ribes hirtellum* (American)

PLANT TYPE: bush

USDA ZONES: 3–8, best in Zones 5 and colder

HEIGHT: 3–5 feet

SOIL: average, loamy garden soil. Will tolerate heavy or sandy soils if well-drained.

LIGHT: full sun or part shade in warmer areas

WATER: regular. Water during driest part of summer.

GROWTH HABIT: straggling bush with upright canes. Some varieties are more spreading.

PROPAGATE BY: cuttings

SPACING: 3–5 feet

YEARS TO BEARING: two (three to four years for full production)

PRUNING: Prune before growth starts when plants are four years old to remove oldest branches. Then remove canes more than two years old each year during dormant season, leaving six to eight canes.

HARVEST: 5–10 pounds per plant. Pick when fully ripe for best flavor; berries are soft when ripe. Wear a leather glove on one hand to hold canes of thorny gooseberries and pick carefully with the free hand. The berries can be kept in the refrigerator for two weeks.

NOTES: Gooseberry plants are illegal to grow in some areas because they host blister rust that kills American white pines; check restrictions in your area and buy only rust-resistant varieties. Gooseberries are self-fertile. They are very cold hardy and will not perform well in warm-winter climates. Gooseberry plants are productive for about fifteen to thirty years. Be aware: gooseberries have thorns, but some varieties are spinier than others. For 5 gallons of wine use about 11 pounds of gooseberries.

BEST USED IN: white table wine and champagne

Kiwi

BOTANICAL NAME: *Actinida* species

PLANT TYPE: vine

USDA ZONES: 5–9 (*Actinidia arguta*, arguta hardy kiwi); 3–9 (*A. kolomitka*, Kolomitka kiwi); 7–9 (*A. deliciosa*, fuzzy kiwi)

HEIGHT: spreading vines whose height depends on the height of the support. Except for the less vigorous Kolomitkas (which trail to 15 feet), kiwis will spread up to 30 feet.

SOIL: well-drained

LIGHT: full sun with wind protection. Kolomikta kiwis require partial shade.

WATER: regular during the growing season

GROWTH HABIT: vigorous vines

PROPAGATE BY: cuttings and grafting

SPACING: 15–20 feet for fuzzy and arguta kiwis; 8 feet for Kolomitka kiwis

YEARS TO BEARING: three to five

PRUNING: At planting, the vines should be cut back to four to five buds. As these grow, select one to be the main stem; stake it to the top of the trellis. On mature female plants, remove shoots that have fruited for three years. Shorten new shoots to three to six buds beyond where fruit last grew. Prune when dormant.

HARVEST: Kolomitka kiwis ripen in August. Other kiwis ripen in October and are picked after the first frost. A mature female arguta or fuzzy kiwi plant will produce at least 25 pounds of fruit. Fuzzy kiwis are picked just as the fruit begins to soften and are ripened off the tree (often in egg cartons). Arguta is picked when the fruit begins to soften. Kolomitka fruit is a bit smaller than arguta and harvest is the same.

NOTES: All species of kiwi need a female and a male to set fruit. Kiwis live to fifty years and longer. Vigorous kiwis require a very strong trellis, 7–8 feet tall. Don't fertilize the first year; afterward, fast-growing kiwis need annual applications of high-nitrogen fertilizer. When making wine, kiwis may be peeled or not, depending on your sensibilities, before they go into the primary fermenter. Nine to ten pounds of kiwi fruit makes a 5-gallon batch of wine.

BEST USED IN: crisp, dry white wine. The wine often retains a hint of kiwi flavor. Strawberry-kiwi is a popular combination.

Loganberry

BOTANICAL NAME: *Rubus loganobaccus*

PLANT TYPE: biennial bramble

USDA ZONES: 5–10

HEIGHT: Loganberries trail 8–10 feet without pruning. Canes should be trimmed to 5–6 feet.

SOIL: deep, well-drained, slightly acid. Mulch is beneficial. If soil is heavy, plant in raised beds.

LIGHT: full sun

WATER: regular through the growing season

GROWTH HABIT: trailing. Perennial roots bear biennial canes. One cultivar is thornless.

SPACING: rows 6 feet apart; plants 2–3 feet apart within rows

PROPAGATE BY: canes that sprout from the roots. Buy certified disease-free plants; those from a friend's garden may carry disease.

YEARS TO BEARING: two

HARVEST: 1–4 quarts per plant; ripens mid- to late-summer

NOTES: Loganberries are a cross of red raspberry and either blackberry or the native Pacific dewberry. Loganberry canes should be trained onto a trellis. Prune new canes to 5½ feet in the dormant season. Remove canes that have fruited when the harvest is done, or in fall or winter. Remove diseased or weak canes and those growing outside the designated bed. Thin the remaining canes to the best five to twelve canes. Loganberries retain their cores when picked. Twelve pounds of loganberries make a 5–gallon batch

BEST USED IN: full-bodied red wine and sweet dessert wines

peach blossoms

Peach

BOTANICAL NAME: *Prunus persica*

PLANT TYPE: tree

USDA ZONES: 5–9

HEIGHT: 12–20 feet. Genetic and grafted dwarfs grow to 5–7 feet. Properly pruned standard peaches are kept at 10–12 feet.

SOIL: well-drained, light, sandy soil that warms up quickly

LIGHT: full sun

WATER: regular during fruit development

GROWTH HABIT: small tree

PROPAGATE BY: grafting

SPACING: standard peaches, 18–20 feet; dwarf 7–8 feet

YEARS TO BEARING: two to four

PRUNING: Peach trees require heavy pruning to produce well. Old wood is pruned away to encourage the new growth on which the fruit grows. Prune lightly for the first five years to encourage fruiting; switch to heavier pruning once trees begin to bear. Fruiting spurs bear for about four years, at which time they should be removed.

THINNING AND HARVEST: 30 to 50 pounds of fruit for a standard-sized peach. Thin clusters to 6 inches apart, one fruit per cluster, in mid-spring when the embryonic fruit is no bigger than a nickel. Pick when fully ripe. Leave stems on. Store peaches on the kitchen counter for three to four days or in the refrigerator in a plastic bag for five to six days. Don't wash the fruit until you're ready to cut it for the fermenter. Remove pits before fermenting as a cracked pit will make the wine bitter.

NOTES: Peaches need a long growing season, mild winters, and 600 to 900 chill hours below 45 degrees Fahrenheit. Peach trees are productive for fifteen to twenty years. Most are self-fertile but will set more fruit with another peach close by. Use little or no fertilizer to discourage overly vigorous growth. These trees are susceptible to peach leaf curl, so choose resistant varieties or commit to spraying lime sulfur at bud-break in late December or early January, and again three weeks later. Dwarf peaches are good candidates for containers; if stored under cover (with sun, but protected from rain, as in an open, south-facing garage) from the onset of dormancy until the blossoms are spent, they will remain disease-free and can be moved to a sunny location for summer. For 5 gallons of wine use about 15 pounds of peaches.

BEST USED IN: dry or off-dry white table wine. Also good for brandy. When fermenting, peaches need yeast nutrients.

Plum

BOTANICAL NAME: *Prunus × domestica* (European plums); *Prunus salicina* (Japanese plums)

PLANT TYPE: small trees

USDA ZONES: 5–9 (European); 6–9 (Japanese); 3–8 (hardy hybrids)

HEIGHT: 15–20 feet. Plum trees can be pruned to 10–15 feet, especially those on semi-dwarfing rootstocks.

SOIL: European plums can grow in heavy soil. Japanese plums do better on lighter, loamy soil. Both prefer a slightly acidic situation and good drainage.

LIGHT: full sun

WATER: moderate yet consistent during the growing season. Certain rootstocks are tolerant of wet soils.

GROWTH HABIT: upright (European plums) or spreading (Japanese plums)

PROPAGATE BY: grafting

SPACING: 15–20 feet

YEARS TO BEARING: four

THINNING AND HARVEST: Thin the fruit of Japanese plums to 5 inches apart when they are no bigger than a dime. European plums rarely need fruit thinning. Expect 2–6 bushels of fruit per tree.

NOTES: Many plums need a pollinizer planted within 40 feet. Japanese plums are very early blooming and may be affected by late frosts. Japanese plums are better in warmer climates; European plums are better for colder climates. European plums don't require much pruning when mature. Japanese plums are bushier and require the branches to be thinned annually. Semi-dwarfing rootstocks keep trees only slightly more compact than a standard-sized tree. Hardy plums that are crosses of Japanese and wild American plums are available for immoderately cold, windy climates. For 5 gallons of wine use about 16 pounds of plums.

BEST USED IN: sherries, ports, cordials, and liqueurs. Some plums result in wines with a spicy taste or odd bite. Plums make a good dry table wine or slightly sweet wine. In Britain, plum "jerkum" is cider made of plums. Plum pits may go in the fermenter as long as they aren't crushed or broken, which can cause bitterness in the wine.

Plum (Early Gold) fruit

Red Raspberry

BOTANICAL NAME: *Rubus idaeus*

PLANT TYPE: biennial bramble. One variety is thornless.

USDA ZONES: 3–8

HEIGHT: Raspberries trail 8–10 feet without pruning. Canes should be trimmed to 5–6 feet

SOIL: deep, well-drained, slightly acid. Mulch is beneficial. If soil is heavy, plant in raised beds.

LIGHT: full sun

WATER: regular through the growing season

GROWTH HABIT: upright/arching. Perennial roots bear biennial canes.

Red raspberries come in single-crop summer-bearers and two-crop fall-bearers.

SPACING: rows 6 feet apart; plants 2–3 feet apart within rows

PROPAGATE BY: canes that sprout from the roots. Buy certified disease-free plants. Plants from a friend's garden may carry disease.

YEARS TO BEARING: two

HARVEST: 1–4 quarts per plant

NOTES: Raspberry canes should be trained onto a trellis. Prune new canes to 5½ feet in the dormant season. For single-crop varieties, remove canes that have fruited when the harvest is done, or in fall or winter. For two-crop varieties, cut last year's canes down by one-third; remove them entirely after the second-year crop. Or cut all canes of two-crop varieties to the ground before spring growth begins for a single late-summer crop. For both types, remove diseased or weak canes and those growing outside the designated bed. Thin the remaining canes to the best five to twelve canes. Raspberry plants are at their peak of productivity for fifteen years. For 5 gallons of wine use about 15 pounds of raspberries.

BEST USED IN: red wine, especially sweet wine

Thornless Brazilian raspberry plants work great in containers.

Rhubarb

BOTANICAL NAME: *Rheum rhabarbarum*

PLANT TYPE: herbaceous perennial

USDA ZONES: 3–9

HEIGHT: 3–4 feet

SOIL: fertile, well-drained, high in organic matter—although rhubarb will grow in most soils

LIGHT: full sun. Partial shade in warmest climates.

WATER: Keep soil barely moist in summer.

GROWTH HABIT: crowns consisting of fleshy rhizomes and buds from which edible petioles (stalks) grow

PROPAGATE BY: division

SPACING: 4 feet

YEARS TO BEARING: Wait until the third year to harvest stalks.

PRUNING: no pruning needed

HARVEST: Pull stalks sideways and upward. Remove when the stalk snaps loose from the base. Don't cut, which leaves behind a stub that can rot.

NOTES: Only the stalks should be fermented (the leaves are toxic). Rhubarb needs some winter chill; it does not do well where summers are hot. Fertilize and keep watered after harvest. A rhubarb plant remains productive for eight to fifteen years. Twelve pounds of rhubarb make a 5-gallon batch of wine.

BEST USED IN: fruity, mild rosé table wine or sweet wine. Rhubarb wine benefits from a year's maturation. Good blended with strawberry and loganberry.

Strawberry

BOTANICAL NAME: *Fragaria × ananassa* (includes June bearers and day neutrals)

PLANT TYPE: perennial groundcover

USDA ZONES: 4–10 (depends on variety); alpine strawberries (*Fragaria vesca*) are exceptionally winter hardy.

HEIGHT: 6–8 inches

SOIL: rich, well-drained, acidic, high in organic matter

LIGHT: full sun

WATER: regular, especially in summer. 1 inch of water a week, more in hot weather. Mulch to keep soil cool and retain water.

GROWTH HABIT: crowns from which leaves, flower stalks, and (usually) above-ground runners grow.

Runners are long stems that "run" along the ground and produce new plants. The new plants root into the soil, grow, and then send out runners of their own.

PROPAGATE BY: new plants that grow at the end of runners; seed. June-bearing plants should be replaced every three to five years, day neutrals every two to three years. They are replaced either by new, rooted plants from the runners or from newly purchased, certified disease-free plants.

PLANTING: Strawberry plantings are managed in two different ways—with runners or without. Runners can either be put to work filling empty

strawberry blossoms

spaces, or they can be removed entirely. June bearers are commonly allowed to grow runners, which are carefully placed to fill in the bed for a solid mat of strawberries. This results in larger crops of smaller berries. Day neutrals are more likely to be kept runner-free, which redirects the plants' energy into larger plants and berries. In early spring, plant your strawberries with the papery growth below the leaf petioles showing, but with all the roots under the soil.

SPACING: 1 foot

YEARS TO BEARING: two

PRUNING: Pick off all flowers from June bearers the first year. For day neutrals, pick off the first flush of flowers.

HARVEST: ½ to 1 pound per plant

NOTES: Strawberries come in two basic types, June bearers and day neutrals. June bearers deliver big berries in June and July. Day neutrals produce a similar-sized crop, only spread out from May or June through first frost. The berries are often not as large as those of June bearers. Most strawberries are self-fertile. In colder areas, the beds are covered with straw in late fall to protect from winter cold; the straw is pulled off the plants and used as a mulch in early spring. Feed June bearers once when growth begins and again after fruiting. Feed day neutrals small amounts of complete organic fertilizer once a month throughout the growing season. Strawberries don't compete well with other plants, so keep the beds scrupulously weed free. Strawberries aren't technically berries; they're aggregate accessory fruits. Now you know. For 5 gallons of wine use about 16 pounds of strawberries.

BEST USED IN: wine, especially sweet wine

Sweet Woodruff

BOTANICAL NAME: *Galium odoratum*

PLANT TYPE: spreading groundcover

USDA ZONES: 4–8

HEIGHT: 1 foot

SOIL: prefers rich soil, but grows in any moist soil within its zone.

LIGHT: shade

WATER: keep moist

GROWTH HABIT: spreads by underground stolons

PROPAGATE BY: division

SPACING: 3 inches

YEARS TO BEARING: whenever the plant is large enough to harvest

PRUNING: can be trimmed during dormancy

HARVEST: snip off fresh new growth in April and May

NOTES: Sweet woodruff can spread quickly through a shady area with rich soil and regular moisture.

BEST USED IN: May wine, in which a dozen sprigs of sweet woodruff are infused in a bottle of Riesling or other young white wine (a young wine is one meant to be consumed immediately, not aged). The wine is then placed in the refrigerator for ten hours to a full day. The beverage is served with fresh fruit, often strawberries. Sugar, champagne, or soda water may also be added. The woodruff brings sweet, spicy overtones to the wine.

CHAPTER 5

Cider

The Lowdown on Apples and Other Cider-Friendly Fruits

A s American as apple pie? Think again. When the colonists planted their orchards they had something else in mind. Two hundred years later, when Johnny Appleseed wandered the wilderness patting pips into feral soil, it wasn't so that hardscrabble homesteaders could enjoy a nice dessert. No, our forebears weren't obsessed with flawless red fruit that could be eaten out of hand; they preferred their apples juiced and fermented. On the frontier, apple trees meant hard cider, the true spirit of the New World. Forget the pastry; the real measure of patriotic purity is as American as apple cider.

Fruit that could be cooked or eaten fresh was certainly a bonus, but what settlers really needed was a drink both tasty and dysentery-free. Water could carry a host of nasty microbes, from *E. coli* to cholera and typhoid. The ethyl alcohol of fermented cider keeps these bowel-destroying pathogens at bay. Other low-alcohol options such as beer and wine were reluctant to adapt to the new continent. European grapes, hops, and barley sulked and succumbed to unfamiliar pests and diseases. Wine made from indigenous grapes was simply unpalatable.

Fortunately, parched sodbusters found the cure to their thirst hanging from the branches of the resilient apple tree. Apples have an amazing adaptive advantage: they produce wildly variable offspring. The trees may be tall or shrubby. Leaf shapes and colors vary, as do branching habits. And the fruit! Sexually propagated apple fruit appears in a seemingly endless array of colors, sizes, textures, time of ripening, and—of course—flavors. We may hope our seed-grown tree will offer up sweet apples with a juicy crunch, but the new fruit are just as likely to be punky, puckery, or taste the way old socks smell.

This genetic scramble makes breeding for specific traits a challenge for pomologists, but it gives the genus a distinct advantage in conquering new lands. If enough seeds are allowed to sprout and grow, a few are likely to possess characteristics that will allow them put down roots in their new home. A hybrid may have a tolerance for sub-zero winters, survive damp soil, or have a natural resistance to local pests. Apples seem limited only by the need for a relatively temperate climate. Although many wildlings will produce "spitters"— fruit unfit for eating—this astringency is a bonus when it comes to cider. Once the apple had settled in North American soil, it began to pursue its own manifest destiny, possibly interbreeding with native crabapples, branching out apace with the country itself.

Throughout the seventeenth and eighteenth centuries, cider was a staple of American life. Everyone, it seems, consumed the mildly alcoholic beverage, even children—who were offered a watered-down version known as ciderkin. Hundreds of thousands of gallons were produced annually. Not only was fermented cider an important part of the pioneer diet, it was a means of preserving a valuable harvest.

For two hundred years, Americans and apples enjoyed a mutually beneficial coexistence: the humans got a healthy drink that delivered a gentle buzz; the trees got a golden era of diversity and the chance to colonize a continent. Then, in the late 1800s, along came Carrie Nation, the Temperance Movement, and eventually Prohibition, and down went the apple trees, literally and figuratively. The apple had become forbidden fruit. As cider fell out of favor, orchards were destroyed and generations of apple diversity were lost. Eventually, in a massive re-marketing, the apple emerged with a sober new image as a nutritious fresh treat and dessert ingredient.

Now cider is back. From 2009 to 2013, US sales of hard cider increased fifty-four percent annually. What's old is new again as we come to re-appreciate fermented apple juice as a drink not only delicious in its own right, but as a natural accompaniment to food of any ethnicity. Could America's thirst for cider eventually outstrip that of the Brits, who have the highest per capita consumption of the drink and turn nearly half of their commercial apples into cider? Craft cideries are springing up across America faster than apple pippins on the frontier. And the trend has not gone unnoticed by hobbyists. Cidermaking lends itself to home production; after all, for centuries it was the stuff of peasants and farmers who didn't have the luxury of fancy equipment or free time.

Freshly pressed, raw apple juice is known as sweet cider in the states. If not chilled or pasteurized, the amber liquid quickly begins to ferment into hard cider. The term "hard cider" is a modern American one. Before the accessibility of refrigeration to slow the fermentation process, virtually any cider more than a few hours old was already "hardening." In European pubs and cafés the drink has always been known simply as cider—the alcohol is implied. This alcohol comes as a waste product of yeasts that naturally accompany apples. So anxious are these yeasts to gobble up the liberated sugars of juiced apples that they are often capable of fermenting all on their own, although cidermakers may choose to replace them with commercial yeasts for better control or to achieve a specific effect. Yeasts will consume all available glucose and then die from lack of further nutrition and from their own byproduct, ethyl alcohol. The sweetest fruit makes the strongest drink, because the yeast has more to work with. Most sweet cider contains just enough natural sugar to result in a spirit of three to six percent alcohol, making it a mildly alcoholic beverage that the ploughman of yesteryear could enjoy with his lunch while keeping his furrows straight in the afternoon.

CIDERMAKING BASICS: COLLECT, CHOP, GRIND, FERMENT

If you're considering adding fruit trees to your garden but wonder whether cider-making might be for you, read on. If you already know how to make cider, but have never had to process fresh apples on your own, this section may be useful as well. If, however, you have attended a club pressing or worked with a local orchard in the past, you may already be familiar with the following. Nevertheless, let's sprint through the basic life of a cider before we get back into the garden!

Cidermaking begins with the collection of apples in the fall. Use only ripe apples. The riper the apple, the more sugar is available for a full-bodied beverage. If the apple seeds are nearly black and the stem snaps when the fruit is lifted, it is ready to be picked. Take care with windfalls. If they aren't split or heavily bruised you may choose to use them; however, although windfalls have long been part of cider, there is a small risk they may pick up botulism toxin from sitting on the ground. Don't allow the fruit to lie longer than necessary—especially on damp ground—as it can acquire this and other unwanted bacteria and mold. As you harvest your fruit, keep in mind that a bushel of apples will yield 2 to 3 gallons of cider. A bushel weighs 48 pounds and has approximately 126 medium-sized apples.

Scabby, unsightly apples that are otherwise sound are born cider apples. Proper apple etiquette dictates that as the fruit is gathered it be piled into a low mound traditionally known as a tump. Tumps are stacked on a tarp or a concrete or wood surface that is reasonably dry and keeps the fruit off the ground. Add more fruit as it becomes available, until the harvest ends or you have enough to stop picking and start pressing. Turn the pile occasionally and discard any fruit showing brown, mushy spots. When it comes to mold and botrytis fungus, one rotten apple really can ruin a whole tump.

Allow the apples to sit for up to ten days to mellow. This softens them for grinding and encourages sugar development. The next step is an important one: you must wash the apples to remove bac-

Talk Cider like a Pro

Biennial: normal-sized harvests every other year alternate with small-to-nonexistent harvests

Bittersharp apples: high in tannin, high in acid

Bittersweet apples: high in tannin, high in sugar

Graft: a scion (shoot) inserted into a rootstock of a plant of the same species to create a clone of the scion. This is the most common way of propagating named apple cultivars.

Lees: sediment in fermenting that settles to the bottom of the fermentation vessel. The lees are usually left behind by one or more rackings

Pip: an apple seed

Pippin: a young apple seedling

Pomace: the pulpy result of mashing, grinding, or milling apples in preparation for pressing

Pomologist: a scientist who studies fruit

Racking: moving a fermented beverage from one container to another using a siphon—not pumping, not pouring. This clarifies the final product by separating the liquid from the sediment.

teria and unwanted strains of yeast on the skin. Nothing but water is required, although—if you like—you can purchase a commercial fruit-and-vegetable wash from the local grocery. Haul a big tub into the yard and fill it halfway with water. Roll the apples in and give them a thorough swish around. As they bob in their bath, give each apple a brisk rub before moving it to the cutting table. If the volume of apples is

Rootstock: the root system and a portion of stem or trunk that makes up the lower part of a grafted plant

Scion: a shoot or bud that makes up the top portion of a graft all growth above the graft should come from the scionwood. More than one variety of apple scion can be grafted onto a single rootstock.

Scratter or scratting mill: a machine that crushes apples or pears in preparation for pressing

Scrumpy: a term applied to high-alcohol cider or perry that can mean 1) an inferior product, 2) a high quality product, 3) a regional product, 4) how you feel after drinking too much high-test cider or perry

Sharp cider apples: low in tannin, high in acid. May also be good culinary apples.

Spur: a short branch specialized for producing fruit

Sweet cider apples: low in tannin, high in acid. May also be good eating apples.

Tump: a low mound of apples being collected for pressing

large, change the water after every batch or two. Discard rotten or severely bruised fruit. For every 20 pounds of apples you'll gain about 5 quarts of cider.

On the cutting table, remove brown spots and quarter the fruit in preparation for pulverizing. Don't remove the cores and pips, as they add flavor. Pulverizing can take the form of mashing, grinding, or milling the fruit so that the press can wring out every sweet drop of future gold-in-a-glass. The pulverizing can be accomplished with a hand-cranked or electric scratting mill, or via a very low-tech plastic bucket and pole. The pole or piece of lumber should be untreated, approximately 5 feet tall, and have a 4- to 5-inch butt. It will act as a pestle to pound the apples into what others call pulp, but you, the cidermaker, will call pomace. Be sure to set the bucket on solid ground so that the pole doesn't burst through the bottom. A stainless steel bucket may be used, but any other metal will react with the acids in apples and give the cider an off taste and color. (Stainless steel is the only metal that should come in contact with your apples and juice from here on out.) You'll know the crushing is done when the pomace is the consistency of wet oatmeal and your frustrations are gone.

Pressing is a bit less primal. In order to squeeze out the essence of the fruit, you need a machine. Presses can take several forms, from the traditional screw press to a car jack. The difference in presses comes down to the way the apple pulp is contained and how pressure is applied. The wood-slatted tub is perhaps the most iconic image of apple pressing. In this form of press, the pomace is placed into a cloth or polyester net bag before being snugged into the tub and juiced. Alternately, apple presses can be frames in which racks are stacked and squeezed. To hold the pomace between racks, you will need to wrap each layer into the aforementioned polyester cloth or mesh.

Apple presses compress the fruit with either a screw mechanism, ratchet, or by hydraulics. Pressing can't be rushed; whenever the output slows to a trickle, the pressure is increased until the flow stops altogether. As soon as the juice hits the air, it will darken due to the oxidization of the tannins that bring dryness to hard cider. But don't fret; fermentation usually restores its golden tone. Taste the juice frequently and, if several varieties of apples are available, adjust the mix to your liking before you proceed with fermentation.

Fermentation can be as simple as putting the fresh juice in an airtight carboy and letting nature take its course. However, nature is unpredictable. The wild yeasts that haunt apples can yield a cider that is less than tasty. Worse, certain bacteria can turn the cider to vinegar. Cider can be made at home in basically the same way as our ancestors did with nothing but nature and personal taste as a guide. The question is—do you feel lucky?

Established cideries build up resident yeasts in their very bones; racks, cloths, vats, even walls eventually come to harbor cider-making yeasts that give the house brands their distinctive flavors. There's no guarantee at the start, however, of what that flavor will be. Given the unpredictability of wild yeasts, and the relative rarity of prime cider-making *Saccharomyces* yeasts, many beginning ciderers rely on commercial yeasts such as champagne, white wine, or English cider yeasts. The fresh juice can be treated with sulfite (in the form of a Campden tablet) to kill or weaken the existing yeasts and harmful bacteria twenty-four hours before the cultured yeast is added. In cider-making's simplest form, pour the juice into a large jug with a narrow neck such as a glass carboy or a polyethylene container. Add the commercial yeast of your choice, and cork the jug with a fermentation airlock.

Before fermentation begins, serious cidermakers check the pH and tannin levels of the juice. Homebrew shops sell pH strips to test for an appropriate level of acid, between 3.2 and 3.8. If the reading is high, add potassium carbonate or another base recommended by the brew shop. If the reading is low, add a small amount of malic acid or other acidity-raising compound and follow directions. The test for tannins is less scientific: Take a taste. If your mouth puckers to the point you need a straw to eat, there is too much tannin, which can be adjusted with gelatin finings. If, on the other hand, the cider is about as exciting as water, it would benefit from powdered grape tannin (and next time, throw some cider or crabapples into the press). Once again, your homebrew proprietor is your best friend.

With the airlock firmly in place, your incipient cider can now bubble along cheerfully. The recommended fermentation temperature is 60 degrees Fahrenheit—give or take five degrees—although slightly lower temperatures only temporarily slow the process and may even improve the quality. The higher the temperature, the faster the fermentation, but too fast is too fast; at temperatures above 70 degrees Fahrenheit the cider risks acquiring an unpleasant chemical taste. Better too slow than too fast; however, very low temperatures for an extended period can lead to fermentation stopping completely and the cider starting to spoil.

See those dregs at the bottom of your fermentation vessel? That sediment, known as the lees, is made up of dead yeast and other impurities. Cider is first "racked off" or separated from its lees when the initial fermentation is complete. Racking off clears the liquid and avoids an off taste from the dying yeast. The first fermentation may last from weeks to months (temperature dependent). When the steady bubbling in the airlock has slowed to a trickle, the cider is ready to be racked off and to begin its second fermentation. The best method for racking off is through simple gravity-based siphoning—not pouring or pumping that can stir up the very sediment you are trying to leave behind. The carboy full of cider sits atop a table; the siphoning tube leads from it to another, empty, carboy on the floor below. Once the racking is complete, the cider is ready for the second stage of fermentation. The second fermentation is a spontaneous process that converts malic acid to lactic acid and carbon dioxide, resulting in a smoother glass.

Allowed to ferment for about six months, sweet apple juice is transformed into a dry beverage with about half the proof of wine. Fermented apples develop acids and tannins that result in a final product which has a taste as independent of apples as wine's is of grapes. Depending on the apples used and the cidermaker's skill, cider can manifest as dark or light, astringent or slightly sweet. It may be naturally or artificially effervescent, or still as the eye of a hurricane. It may taste citrusy, smoky, musty, or like molasses. The average five-percent alcohol content of cider can be raised if additional sweetener is added to the fermenting juice to keep the yeast alive and digesting. However, if you boost the alcohol to between eight and fourteen percent, your cider is now technically wine.

CIDER APPLES

Cider is, of course, all about apples. Yeasts and additives have an effect on the outcome of a given batch, but apples are the real flavor maker. Apples contribute four basic characteristics to cider and are classified according to these qualities as sharp, sweet, bittersharp, or bittersweet. Popular proportions are one part sweet, one part bittersweet, and one part sharp or bittersharp apples. There are hundreds of varieties of cider apples that offer a wide variety of tantalizing flavors, complex aromatics, and tangy tannins. Cider makers carefully mix the juices for a drink that makes their mouths water. As you become a more experienced ciderer, you will discover your own personal ratios.

To create the absolute best in cider alchemy, choose your apple varieties thoughtfully. Of all the tree fruits that can be grown in temperate backyards, apples are the most gratifying. Their only silver bullet is warm climates that don't give the apples at least 500 to 1,000 hours of temperatures below 45 degrees Fahrenheit. Even then, the fruit may suffer mainly in color. At the other extreme, apples resent bitterly frigid winters, so are grown on certain hardy rootstocks in the upper Midwest and Alaska.

This doesn't mean that every apple cultivar grows equally well in every region. For instance, Cox's Orange Pippin does not perform well in climates where summer highs are consistently above 86 degrees

Fahrenheit. Braeburn drops its fruit in hot weather. Granny Smith—as is appropriate for a granny—doesn't like the cold. Before choosing your varieties, check with local nurseries, orchardists, and your County Extension service for recommendations.

Now comes the tricky part. A hundred years ago, cider fell out of favor. From that point on apple varieties were selected for fruit that could be eaten fresh or cooked, not for fermenting. Although cider is making a precipitous comeback, cider trees are much less well-known and available than those for culinary uses. The best selection is by mail-order through specialty nurseries. Local extension agents may not be familiar with cider cultivars. If you intend to take your cider-making seriously and plant a Brown Snout or Roxbury Russet, you may be the first in your area to write a review.

Cider quality inevitably depends on the type of apple used. Culinary apples can—and commonly are—used for cider. However, making cider without any cider apples is like making pickles without dill; it's perfectly possible, but not what nature intended. Try to include at least some true cider varieties or even crabapples. The high tannin level of traditional cider apples improves the flavor, body, and complexity of the product. A mix of cider and dessert apples provides a balance of tannins, acids, sweetness, and aromatics, which results in a well-rounded beverage. All apples can be part of a fine cider; the key is in knowing the characteristics of each and recognizing its

Apples are one of the most familiar and versatile fruits to grow.

place in blends, or if it's one of the few that can stand alone as a single-variety cider.

Most cider apples are no one's first choice for eating raw. They are often tart, marred, and misshapen. To the human palate, cider apples often taste unripe. Their potential is realized only at decanting. Identifying the best cider varieties is an impossible task. Favored cider cultivars come and go with the fickleness of fashion. Even more confusing, from nation to nation and person to person there is no consensus on what makes a perfect cider. What is certain is that the very lack of tannins and potent acids that make modern apples so tasty is also what makes insipid cider. Dessert apples can add sparkle and depth, but a quality cider demands at least a bit of the cantankerous kick of otherwise unappetizing apples.

The tannin required for spunky cider can come from vintage or vintage-quality cider apples, or from crabapples. Crabapples lack the complexity of their bigger cider cousins but are bristling with astringency. Crabapples such as Dolgo and Evereste™ are easy to grow, cold hardy, disease resistant, and beautiful in flower and fruit. Most crabs grow to only 10 to 20 feet, thus fitting nicely into ornamental gardens. The fruit comes in shades of yellow, coral, and red. Blossoms are white or pink, and the foliage may be green or bronze. Crabapples are also excellent pollinizers for other apple trees. Before committing to a crab, be sure it's a variety both hardy to your region and resistant to local diseases.

If you're serious about cider, or know someone who is, one or more cider trees can be a good investment. Kingston Black is celebrated as the finest vintage apple for single-variety cider. Its small, maroon fruit results in a spicy, full-bodied drink. On the downside, the tree can be difficult to cultivate. It is a somewhat weak grower and is susceptible to scab and canker. In addition, its fruiting is unreliable. The harvest may be large, or nearly nonexistent.

There are a few other apples with the required balance of tannin and acid to stand alone as a single-variety cider. These include Foxwhelp, Stoke Red, Golden Russet, Roxbury Russet, and Ribston Pippin. However, many fine varieties are available that mix well with modern culinary apples. Although today's fresh-eating apples taste sweet, this "sweetness" is often due to meager malic acid rather than an excess of sugars. Incorporating them with apples labeled sharp, bittersharp, or even bittersweet can be a recipe for a first-rate cider.

Rootstocks

The roots that will support your tree are as important a consideration as the variety that grows atop them. Named apple varieties are almost exclusively grafted onto rootstocks (the exception being trees grown from cuttings, therefore on their own roots; seed-grown trees are genetically unique from their parents—thus no longer a named variety). The rootstock determines how big the tree will grow, how hardy it will be, and if it is particularly canny at surviving drought or damp. If possible, buy your plant from a local nursery that is familiar with local conditions. Depending on the rootstock, a cider apple can grow as a dwarf (less than 9 feet), semi-dwarf (up to 14 feet), or standard (15 feet or more).

Apple trees grown from seeds or cuttings—not grafted to specialized rootstocks—tend to grow to standard size. The amount of room you have for apple trees, or the number of apple trees you desire, will affect your choice of rootstock. Each standard tree requires 20 feet apart from its neighbors in all directions; large semi-dwarfs need 16 to 18 feet to each side, smaller semi-dwarfs 12 to 14 feet, and dwarf trees just 5 to 7 feet. Dwarf trees produce full-size apples and relatively large crops (about a bushel) on a tiny tree; however, because the same root system that stunts them also gives them a poor grip on the ground, these Lilliputians generally need permanent staking. While dreaming of growing your own apples, remember that two varieties (or more!) are better than one for thorough pollination. Just be sure that at least two of them are listed as pollinizers. Of course, existing apple trees growing nearby can also provide the necessary pollen.

Williams Pride apple flowers

Apple Varieties to Consider

Bramley's Seedling: sharp. Vigorous, hardy, heavy crops. Excellent culinary apple, as well as for cider. A fair keeper. Ripens October.

Brown Snout: bittersweet. Self-fertile. Slow to produce fruit. Susceptible to fireblight. Ripens November.

Brown's Apple: sharp. Vintage. Scab resistant. Slow to produce fruit. Ripens October.

Campfield: bittersweet. Vigorous. Single-variety cider or blend with Harrison. Long keeper. Ripens October.

Cap of Liberty: bittersharp. Vintage. Best blended with a sweet or bittersweet variety. Ripens September.

Chisel Jersey: bittersweet. Vigorous. Produces heavy crops. Ripens November.

Dabinett: bittersweet. Vintage. Weak grower. Good crops. Aromatic. Excellent blender. Ripens November.

Foxwhelp: bittersharp. Vintage. Can be used for a single-variety cider or blended. Ripens September.

Golden Russet: sweet. Aromatic. Can be used for a single-variety cider or blended. Can be eaten fresh. Ripens October to November.

Graniwinkle: sweet. Vigorous. Large crops. Usually mixed with Harrison. Can be eaten fresh. Ripens September.

Grimes Golden: sweet. Aromatic. Good culinary apple. Somewhat resistant to fireblight and cedar apple rust. Ripens September to October.

Harrison: bittersharp. Vigorous. Large crops. Classically mixed with Graniwinkle. Stores into spring. Ripens October.

Harry Masters' Jersey: bittersweet. Vintage. Good crops. Ripens October to November.

Kingston Black: bittersharp. Vintage. The classic cider apple. Can be used for a single variety cider. Small tree that can be difficult to grow. Ripens late September.

Major: bittersweet. Vintage. Vigorous. Ripens late September.

Medaille D'Or: bittersweet. Vigorous. Makes a high-alcohol cider. Ripens September.

Michelin: bittersweet. Excellent blender. Most widely grown cider apple in England. Ripens October to November.

Redstreak: bittersharp. Vintage. Esteemed in England in the 1600s. Ripens November.

Ribston Pippin: sharp. Aromatic. Hardy. Also a dessert apple. Can be used for a single-variety cider. Ripens September to October.

Roxbury Russet: sweet. Aromatic. Possibly the oldest named apple variety in America. Can also be cooked or eaten fresh. Good keeper. Good for cold climates. Ripens October.

Smith's Cider: sweet. Vigorous, bears early, fruits heavily. Can be eaten fresh. Low chill requirement makes it a good choice for warmer areas. Ripens November.

Stoke Red: bittersharp. Vintage. Vigorous, fruits heavily, but slow to bear. Very resistant to scab. Ripens late November.

Sweet Coppin: pure sweet. Vintage. Prone to mildew. Biennial bearer. Ripens October to November.

Tom Putt: sharp. Also culinary. Vigorous, bears early and reliably. Some resistance to scab. Ripens late August to September.

Yarlington Mill: bittersweet. Vintage. Vigorous. High yields. Ripens October to November.

PLANTING YOUR CIDER ORCHARD

Now that you've selected your cider trees, it's time to plant your orchard. Whether you're planting one tree or many, plan your orchard proudly; you're part of the great American apple tradition, and the tree you plant today—if luck and health are on its side—may provide apples to generations of cider makers.

Liberty apples

One thing is critical when deciding where to site your orchard: Apple trees need sun. Too much shade aids disease and thwarts ripening. Set trees out of prevailing winds, which can disrupt pollination. Frost pockets can damage blossoms or fruit in colder climates. Although apples prefer to grow in rich loam, they can withstand less-than-perfect soil. Dwarf trees, with their brittle rootstocks, benefit most from good soil. Dig a hole deep enough that the top of the rootball sits level with the top of the ground or slightly lower where winters are harsh. Fill the hole halfway up the rootball with native soil, and add water to the very top. A thorough soaking is important not just for hydration but for settling the soil and filling air pockets. Once the water has soaked in, finish filling the hole, leaving a shallow depression with a raised outer lip around the top of the sunken rootball to act as a bowl for future soakings. For the first two years after the tree is planted, don't allow the soil around the rootball to dry out. Neither should it be perpetually soggy.

The generally accepted time for setting out fruit trees is in the dormant season, November through January, and this is when bareroot trees are available. However, container-grown trees can be planted in any but the hottest weather. Stake trees only if they are top-heavy or feel loose. Plan to remove the stake as soon as the tree is stable. Dwarf trees are an exception to this rule: they'll need a permanent stake, so take the extra time and expense to install an attractive stake that will last for at least twenty years. Finally, if mice, rabbits, or bucks rubbing velvet from their antlers are a problem, wrap the trunk in hardware cloth or fine-wire fencing from the ground to the first branch. Deer love to eat apple leaves; in a few minutes a marauding doe can strip and kill (or severely set back) a young tree. If there are deer in your neighborhood (or other hungry herbivores), enclose the entire tree in a cage made of stock fencing until the sapling is big enough to survive their grazing.

Grafting

Apple trees do not come true from seed. Quite the opposite. An apple pippin, or seed-grown tree, delights in asserting its independence. It may be the next great pomological sensation or, more likely, a dud. What it will not be is a clone of its parents. For that, an apple tree must be grafted. Grafting is a matter of attaching scionwood, or branches of the desired tree, to a rootstock, the roots and trunk of a donor apple—possibly the very wildlings just discussed.

Grafting has several advantages. For one, it allows the creation of dwarf and semi-dwarf trees. Full-size, "standard" apple trees tend to grow to 20 feet or more, making harvesting their fruit a feat for monkeys. A standard-sized tree grown from seed can take up to ten years to bear fruit, while a grafted dwarf or semi-dwarf tree can yield fruit in as little as two

Grafting apple trees

to four years. Virus-free, dwarfing apple rootstocks such as the EMLA series or the diminutive Budagovsky 9 result in trees 6 to 16 feet tall. Rootstocks such as these are readily available online and from mail-order fruit nurseries.

Grafting is not difficult. With good instruction, a few inexpensive tools, and a half-dozen rootstocks, the novice has a decent chance of getting at least one healthy tree. With a little experience and access to scionwood from donor trees, you can beget an entire orchard for the price of the rootstocks—often the price of a single nursery tree. The home grafter can propagate vintage varieties and even experiment with multiple grafts on one stock, thereby creating a custom combo tree. Imagine growing a full range of cider varieties on a single tree!

Grafting is best done at the very end of the dormant season, just before or as the buds begin to swell. Your most important tool will be a very sharp knife. Specialty grafting knives are not terribly expensive, but a finely honed pocketknife or utility knife can also be used. Since you'll be pulling that sharp blade toward your fingers with a good deal of force, leather gloves are a wise precaution. Wide rubber bands or special grafting bands that disintegrate with age are needed to hold your newly inserted scion in the stock. Wax can also be used to seal certain types of grafts and hold the parts in place.

A graft can be made from either a branch or a bud inserted in an understock. Both branch and bud grafts can be accomplished in several different ways. Apple trees can be propagated by any of these methods, but one of the most common and straightforward is the whip-and-tongue graft. The goal, as with all grafting, is to line up the cambium layer of the two pieces. Located just beneath the bark, the cambium produces the cells that transport water, sugar, and nutrients from the roots to the crown and through every living part of the plant.

Before cutting the scionwood, be sure you know which end is up, and keep it that way. An upside-down scion is a dead scion. Line up the scion alongside the stock to find the point where they are as close to the same dimension as possible. At this point, cut a 7-inch piece of scionwood using pruning shears sterilized with rubbing alcohol. Two inches at the butt end will be whittled to attach to the stock. Likewise, trim the stock 2 inches higher than where it will join the scion to allow for whittling.

Next, use the knife to cut matching diagonal slices 1½ inches long from both the stock and scion. They must match up like two parts of the same stick as this is where the two will be joined. To hold them together securely, each piece gets a second cut, this one from the freshly cut end directly upward into the woody center about ¾ inch deep. On the stock, this should be performed one-third of the way up the fresh cut; on the scion, one-third of the way down. This creates a "tongue" on each end which, when slipped together, gives the graft union added support. A cut rubber band is then wound tightly round and round the graft and tied off. Pot your new apple tree and keep it in a sheltered spot such as a screened porch for the first few weeks to keep it from drying out. It can then be moved outdoors to a sheltered garden spot, out of direct sun and strong wind until the graft fuses. If the tree leafs out robustly the following spring, it's ready for a permanent spot in the garden.

Pruning

Proper pruning will reward you with well-placed branches, good crops of large fruit, and easy harvesting. The goal is to create a strong system of branches angled widely to the trunk and spaced for optimum light and air circulation. Pruning can be done in winter and summer. Each season has its own reasons and results. Winter pruning removes diseased or poorly placed branches; it concentrates the stored energy of the tree in the remaining wood, causing a rush of spring growth that often includes strongly vertical water sprouts from the branches and suckers from the roots. Suckers and water sprouts should be removed flush with their point of origin—pulled off along with their growth points, if it can be done without tearing the surrounding bark. Rub off any growth below the graft as soon as it appears.

Summer pruning has the opposite effect. Removing branches with leaves removes energy, giving the tree less impetus

Pruning is essential for plant health.

to unbridled growth. August is not only the month for trimming trees while triggering fewer water sprouts, it's also the time to create spurs, beneficial stubby branches whose sole purpose is to produce fruit. Many varieties of apple will form spurs where new growth coming from the trunk or main branches is cut back to three leaves. Finally, badly placed or crossing branches and those with narrow crotch angles can be removed in summer rather than winter; in the summer, it's just more difficult to see the overall structure of the tree with all those leaves in the way. Prune newly planted trees lightly for their first few years. Forget about fruit and concentrate on a strong structure, which will determine the shape of the tree all its life. Remove droopy and weak branches. Retain strong branches that emerge in an even spiral up the tree. On semi-dwarfs and standards, slowly remove lower branches as the tree grows to leave a comfortable space beneath the eventual mature tree for mowing or—on standards—a tire swing.

Depending on the rootstock and the cultivar, a new apple tree will begin to set fruit in two to ten years. The number of years the tree will be productive also varies. Shallow and noncompetitive dwarf rootstocks may produce for only twenty years, while standard trees on their own roots can persevere for a hundred years, as proved by old farmyards across the nation.

Thinning

After two or more years of patient waiting, watering, pruning, and protecting, at last your apple is expectant with embryonic fruit. Now, pinch them off. Well, not all of them. This important orchard chore is known as thinning. Thinning picks up where nature leaves off. Apple trees shed some of their immature fruit in early summer in a natural process known as June drop. Homeowners can finish the job by removing all but the healthiest little apple in every 5 or 6 inches. Thinning prevents small or weak branches from breaking under heavy fruit loads. It also inhibits biennial bearing, a condition where a tree over-produces one year, then takes the next year off. Although a thinned tree produces less fruit, the remaining apples are larger and healthier. Thinning should be done before the tiny apples grow larger than a nickel.

Harvesting

You'll know the cider days of summer (or fall) are here when the apples begin to tumble from the trees without the aid of wind, weather, or romping neighborhood children. As a test, lift an apple, putting pressure against the seam of the stem and, if necessary, giving a gentle twist. Ripe fruit will snap off in your hand; unripe fruit will resist. A properly picked apple will come away with the near end of the stem still attached. Pack or stack the fruit gently; a bruised apple is a bad apple, and we all know what one of those can do.

Apple trees and their fruit are beset by scab, powdery mildew, fireblight, canker, codling moth, and apple maggot. That's the bad news. The good news is you're growing cider apples; short of actual rot in the fruit, you needn't worry about how they look. Scab, mildew, even insect blotches don't affect these stalwarts' use in cider. If parts of the fruit are brown and mushy, either throw away the entire fruit or cut out the damaged areas which can host toxin-producing mold. When it comes to insects, the hard truth about hard cider is this: the occasional pressed larva is pretty much inevitable. Do as our pioneer ancestors did—squeeze your apples boldly and try not to think about it.

CIDER BEYOND APPLES

Cider needn't be about apples alone; other homegrown flavors can lend a refreshing— or interesting—twist to this otherwise single-minded beverage. Berries and other fruits are natural cider associates. Practically any berry you can grow in your backyard makes a nice addition to cider. Often-used berries include raspberries, strawberries, blueberries, blackberries, and black currants. These and other juicy fruits such as peaches, cherries, kiwi, pears, and quince are juiced and fermented separately, then added to the fermented apple juice. Drier fruits and herbs can be steeped in the cider at the first racking, after the liquid has been siphoned from one fermentation vessel to another.

THE PERFECT CIDER GARDEN

In spring, when the apple trees bloom, our fancy turns lightly to thoughts of cider. Sit back and crack open a bottle of liquid gold, lovingly put away the previous fall. Let the birds sing; let the bees drone from blossom to blossom pollinating the coming harvest. What could be finer than to relax beneath your trees and enjoy the labors of your fruit?

A 20-foot square of backyard can hold one standard-sized apple tree, two semi-dwarfs, or four fully dwarf trees. Three dwarf apple trees leave room for another type of fruit tree to round out your cider-making possibilities. Depending on your zone, consider a dwarf peach, quince, or sour cherry—great cider additions all.

Peach trees (*Prunus persica*) demand a mild climate: they need at least five frost-free months and a chilling period of 600 to 900 hours below 45 degrees Fahrenheit. Where springs are particularly cold and damp, peaches suffer poor pollination and peach leaf curl. Many cultivars are self-fertile. Dwarf peaches of 5 to 7 feet are available.

Fruiting quinces (*Cydonia oblonga*— not the more common flowering quince, *Chaenomeles* species) are attractive all-season trees that, at less than 15 feet, fit neatly into a home landscape. Quince fruit may be mistaken for oversized, downy-skinned pears. Although now somewhat rare, quinces were common in colonial orchards. They produce heavy crops, but the fruit is slow to ripen, often into November. Gardeners in cold climates may find their quince fruit can't beat the first killing frost. Otherwise, the trees themselves are hardy into Zone 5. Quinces are self-fertile; you only need one.

Sour cherry trees (*Prunus cerasus*) are the most cold-resistant of the three. They can survive in Zone 4, although they might suffer deadly trunk damage in the coldest regions. Blossoms can be nipped by late frosts. Sour cherries perform best where winters are frosty and summers are mild to moderate. The trees are natu-rally small, to 20 feet tall with a spread as wide. Some sour cherry varieties are sweet enough to be enjoyed fresh. Sour cherries are self-fertile.

Your three dwarf apples trees should—in two or three years—present you with enough fruit for about 9 gallons of fresh cider. If your climate is mild and your soil reasonably friable, why not try the king of cider apples, Kingston Black? A bushel of bittersharp Kingston Blacks will enhance any cider. For your second tree, consider a bittersweet such as a vigorous and

Growing Apple Trees from Seed

Apple trees can often be grown from pomace via benign neglect. Spread the seedy pomace on loose, fallow soil in the sun, keep it damp, and wait for the pips to pop. In the dormant season, move the seedlings to pots to grow in, or—if they aren't crowded—leave them in place until they're at least 18 inches tall, and then move them to their permanent homes. There are two things to keep in mind when growing apple trees from seed. First, these will be big trees (up to 30 feet). Second, you never know what you're going to get with sexually propa-gated apples; they're likely to be poor culinary fruit, but fine for cider. That's the fun of it! You've engaged in a botanical treasure hunt. You are the proud owner of apple trees found nowhere else in the world. If one of your pippins turns out to be a genetic gem, you can graft it and share it with friends—even patent it. These trees are yours alone, and they are unique.

high-alcohol Medaille D'Or, or an equally vigorous Major. Your third tree can round out your cider triumvirate with a sweet such as a Grimes Golden, which is aromatic and doubles as a culinary apple, or a Roxbury Russet with the same characteristics but also stores well and thrives in colder climates. In place of a sweet cider apple, you can plant your favorite culinary apple—perhaps a Cox's Orange Pippin, Jonagold, Liberty, Honeycrisp, Ashmead's Kernel, or whatever suits your fancy. Just be sure it ripens around the same time as its orchard-mates; this not only makes for convenient pressing but helps ensure pollination because apples that ripen together tend to bloom together.

Now for some berries. A 20-foot row of raspberries, blackberries, or jostaberries will further expand your cider repertoire. If your orchard comprises dwarf trees, you might even consider growing a row of each along the east, south, and west borders of your garden. After all, raspberries and blackberries are a wonderful summer treat that are at their best picked out of your own backyard. Jostaberries (pronounced YOS-ta-berries: in your best pseudo-Swedish accent say, "Yah, yostaberries!") can also be eaten fresh, but are equally good turned to jelly—or included in cider.

Raspberries come in two distinct types: single-crop summer bearers and dual-crop summer/fall bearers. In summer-bearing raspberries, as well as the upright-growing blackberries, canes that produced last year

are pruned out the following fall or winter. This means careful pruning since this year's fruiting canes are growing amongst last year's dead canes. The dual-crop raspberries are different. The easiest way to deal with these is to simply cut all the canes down in the winter before the new fruiting canes appear in spring; this gives one crop of berries in late summer. Or they can be pruned like summer bearers (removing last year's dead canes) and the live canes allowed to grow for two smaller crops: Summer and early fall. Upright blackberries, in addition to having their old canes removed, benefit from having their new canes tipped back by 3 to 6 inches when the canes reach 3 to 4 feet. Otherwise, the main requirements of blackberries and raspberries are well-drained soil, sun, winter chill, and a Zone 3 to 8 setting for raspberries, Zone 5 to 9 for blackberries.

Jostaberry is a cross of black currant and gooseberry. They are easy to grow, thornless, and generous with their crops. The attractive bushes grow to 5 to 6 feet and prefer Zone 5 and colder. A single plant can gift its grower with 8 to 12 pounds of fruit. Check your local regulations before setting your heart on jostaberries or currants; they are prohibited in some states because they host blister rust, which decimates native white pine.

Refer to the Wine chapter for additional notes and instruction on:

- Blackberry
- Blueberry
- Cherry
- Cranberry
- Elderberry
- Pear
- Pumpkin
- Red Raspberries

What to Do with Pomace

Waste not, want not, and apple pomace is much too good to waste. You can go the traditional route and feed it to your pigs or livestock (though it is too rich for horses). To do this, mix small amounts of pomace with large amounts of hay or other forage. The crumbly mash also makes a luxurious mulch for acid-loving plants such as rhododendrons—if you don't mind its brief, though heady, wine aroma and volunteer seedlings. If nothing else, pomace is an excellent compost ingredient. However, if you're daring, and have an acre to fill, consider growing new trees from the leftover apple seed.

Crabapple

BOTANICAL NAME: *Malus domestica*

PLANT TYPE: tree

USDA ZONES: 4–9

HEIGHT: 20–30 feet standard; 12–15 feet semi-dwarf; 7–10 feet dwarf

SOIL: well-drained, acidic soil

LIGHT: full sun

WATER: regular during fruit development

GROWTH HABIT: tall, spreading

PROPAGATE BY: grafting, seed. (Apple trees do not come true from seed.)

YEARS TO BEARING: standard 5–8; semi-dwarf 3–5; dwarf 2–4

PRUNING: In the early years, concentrate on building a strong scaffolding. Retain strong branches that emerge in an even spiral up the tree. Choose one tall, strong, upright to serve as the central leader and remove any other tall branches that might want to challenge it. When the tree is mature, cut back the central leader to the highest branch coming off of it, around 7 feet from the ground. Remove dead, diseased, weak, and crossing branches, and those growing into the center of the tree. This encourages new wood with new fruiting spurs.

THINNING AND HARVEST: Thin to one fruit every 5–6 inches. At harvest, a dwarf tree can yield 30–60 pounds of fruit; a standard, 300–800 pounds. To see if fruit is ripe, cut open an apple and check for dark brown seeds.

NOTES: Apple trees can live 60 to 140 years. Two varieties are needed for cross-pollination. Dwarf and semi-dwarf trees are often more suited to home orchards. Apples can be grown as an espalier. For 5 gallons of cider use about 20 pounds of apples.

BEST USED IN: sweet and hard ciders, wine, and liqueurs. Apple wine tends to be sweet and buttery.

Jostaberry

BOTANICAL NAME: *Ribes × nidigrolaria*

PLANT TYPE: small shrubs

USDA ZONES: 3–8, best in Zones 5 and colder

HEIGHT: 7 feet

SOIL: average, loamy garden soil. Will tolerate heavy or sandy soils

LIGHT: full sun

WATER: regular

GROWTH HABIT: upright, thornless, branching canes

PROPAGATE BY: cuttings

SPACING: 5 feet

YEARS TO BEARING: two (three to four years for full production)

PRUNING: Prune before growth starts when plants are four years old to remove oldest branches. Then remove canes more than two years old each year during dormant season, leaving six to eight canes.

HARVEST: 3–10 pounds per plant. Pick when fully ripe for best flavor; jostaberries turn nearly black when they are ripe. They ripen individually within clusters, so must be picked individually. Avoid washing, which can cause them to spoil quickly.

NOTES: A thornless cross between a black currant and a gooseberry, jostaberries are vigorous and healthy plants that require little pruning. Self-fertile.

BEST USED IN: cider, wine

Quince

BOTANICAL NAME: *Cydonia oblonga*

PLANT TYPE: small tree

USDA ZONES: 5–9

HEIGHT: 10–25 feet

SOIL: well-drained

LIGHT: full sun

WATER: moderate to moist

GROWTH HABIT: A quince can be trained to a single trunk as a small tree or grown as a multiple-stemmed bush.

PROPAGATE BY: cuttings, layering, seed

SPACING: 20 feet

YEARS TO BEARING: five

PRUNING: Quinces need little pruning.

HARVEST: 30 pounds per tree at seven to eight years. The fruit turns yellow when ripe.

NOTES: Quince fruit resemble large, furry pears. Fruit may not ripen until November and can be lost to freezing in colder regions. Quinces are self-fertile. Store fruit in a single layer in trays. Keep in a cool, dry place where they should hold for up to three months.

BEST USED IN: cider, wine

A Very Scrumpy Sidebar

Who doesn't like the word *scrumpy?* Try saying it aloud—see what I mean? *Scrumpy* is a perfect cross between "scruffy" and "grumpy." It's how I feel around 5 p.m. after a day hunched over my keyboard, still wearing the clothes I slept in. But then I pop open a bottle of cider (or better yet, scrumpy!) and suddenly, it's all good. So let's get our scrumpy on.

Apple or pear scrumpy is strong cider. It's potent, puckery, and usually cloudy. The term—which comes from windfall apples, known by rustic types as scrumps—is otherwise open to interpretation. For some, it means cider that hasn't finished its final fermentation; for others it's a heady drink allowed to mature for an extended period. Still others view scrumpy as a local product, indigenous to certain cider-producing regions of England. In general, scrumpy has an alcohol content of seven to eight percent, but it can run as high as fifteen percent. Maybe "scrumpy" is just how you feel the morning after indulging in too many glasses of fifteen-percent cider.

CHAPTER 6

A Little Bit of Perry

A Chapter for Pears

P ity the pear. It gets no respect. Even our lexicon is against it. A pear a day will not keep the doctor away. You will never travel to the "Big Pear" and make it there. Did you upset the pear cart? Find someone who cares. Sadly, there's no denying that the pear is the apple of no one's eye.

So, rather than lumping pears with apples in the cider chapter, let's give them the dignity of their own chapter—even if it is somewhat brief. In it we will eschew the dismissively derivative "pear cider" in favor of the drink's rightful name: perry. For a few short pages we will celebrate this sadly slighted fruit and the alcohol derived from it. How do you like them pears?

GETTING TO KNOW PERRY

Perhaps the easiest way to categorize perry is by considering what it's not: It's not wine, although pears do make a fine wine. It's also not hard cider; cider is made primarily of apples and tends to have a more robust, earthier flavor. So, why all the confusion between cider and perry? It's a matter of marketing. Rather than introduce Americans and young consumers to the unfamiliar (though ancient) moniker "perry," the architects of euphemism have

determined that the beverage would go down smoother as "pear cider."

Owing to the low acidity of pears, perry tends to be sweeter, smoother, and more delicate than cider. While it's true that the addition of pear juice can mellow and refine any hard cider, this does not make it perry. Real perry is 100 percent pressed perry pears. A traditional and excellent perry is actually more achievable by you, the home-owner with one good perry tree, than it is for corporations whose drinks are derived primarily from general-purpose pears or a mix of pear and apple. Perry requires a little study and a lot of attention to detail. That said, perry-making isn't complicated. A bit of effort and you too will understand why wealthy landowners of yesteryear gave cider to the hired help and saved the perry for themselves.

If cider received little attention in the twentieth century, perry practically

Pears are one of the tastiest and most overlooked fruits in brewing.

disappeared. It was never standard fare in the United States, but even in perry-producing England the drink went into decline as orchards were lost to fire blight and changing fortunes. Migrant pickers moved to more lucrative, less strenuous work. In just a few decades, dessert pears had replaced many perry orchards. Left with mostly sweet pears to work with, aspiring perry producers found their fledgling beverages disappointingly insipid. The near extinction of perry has created certain challenges to modern perry production. For one, true perry pears, which make all the difference, are now hard to come by. For another, cider is simply more familiar to consumers. There's no denying that perry has been sadly neglected. But things are looking up for perry. Thanks to modern fermentation artists, the drink is making a comeback—even if one somewhat open to interpretation and experimentation.

Perry at Home

Another challenge for perriers is the nature of pears themselves. There are several important differences between pears and apples. For one, perry pears are less acidic than cider apples. They have higher levels of sugar including unfermentable sorbitol, which gives perry a sweeter finish than cider as well as a slight laxative effect. Pear tannins also differ from those in apples; they dispose fermenting perry to develop an unappetizing haze that must be dealt with, and even then perry is prone to cloudiness. The interference from this tannin haze is one reason homemade perry is generally a flat beverage; although it might be made to sparkle with a bit of added sugar at bottling, if you're feeling pétillant. Finally, pears have characteristics that can lead to sluggish fermentation. This final failing is often solved by the addition of ammonium sulfate to the juice. If you understand the peculiarities of pears, take a bit of care, and add a few commercially purchased preparations, you can process perfectly passable—even profound—perry.

The home perrier must understand yet another important distinction between the fermenting of pears versus apples: Pears require patience. First, pears need time to mature after harvest. This takes from two days to a full week (sometimes longer), depending on the variety. Second, pear pomace needs a rest between crushing and pressing. After it has been thoroughly mashed, pear pulp should sit for twenty-four hours to allow some resident tannins to dissipate. This step is crucial for reducing cloudiness in the final product.

From harvest to table, a well-made batch of perry can take ten months. Is it worth the wait? Just ask Napoleon, a perry connoisseur who once crowned the beverage "the champagne of Britain." "Oui, oui!" he would say if he were alive. No doubt Napoleon knew to pick his pears when they snapped neatly at the stem joint when lifted and slightly twisted. Follow his example. Treat your pears with respect: a fruit roughly dropped into a box, an incautious fingernail piercing the skin, and spoilage is imminent.

Pear picking can be done safely—with feet firmly on the ground—via an old fashioned fruit picker on a stick. A fruit picker consists of a circular metal tool with "fingers" used to pry the fruit loose, and a bag hanging beneath to catch the fruit as it falls. The metal picker is fixed atop a pole, such as

bamboo, in the length of your choice. Once the pears are harvested, they (and you) are ready for a rest. As previously mentioned, pears will need several days if not a week or two to mellow. Store them in a single layer at room temperature. If pears are used too soon, they will add little of their characteristic flavor to your perry. If they sit too long, you will find they have stealthily rotted from the core outward and are destined for the compost rather than the carboy.

Twenty pounds of pears yield about a gallon of perry. If you plan to go au natural with your fermentation, don't wash the pears, which will remove wild yeasts. If you'll be using commercial yeast (which is recommended), go ahead and rinse your pears. They are now ready for pulping. Pears can be pulped in the same way as apples (see p. 142), or, if they're suitably soft, the fruit can be squished into sauce with anything from a potato masher to a small, barefoot child. For a modest batch of perry, you can crush your pears in a juicer or blender. Better yet, invest in a fruit pulper that attaches to the end of an electric drill and chops the fruit with a blade. After crushing, the pulp—more accurately known as pomace—needs another rest. Cover the pomace with cheesecloth or a towel to thwart fruit flies, and leave it in a cool place overnight to oxidize out excess tannins. This step is important for reducing elements that cloud perry and slow fermentation, risking a microbial bloom. The next morning, the pomace is ready for pressing (see page 137). Beware:

mashed pears are slippery and thus trickier to press than apples; unless care is taken, a post-pear-pressing kitchen may look like the scene of a food fight.

Fermenting the Juice

In the days of old-fashioned perry-making, the juice was poured into wooden barrels and left to the mercy of wild yeasts. The same process can be used by you, the home perrier, with the substitution of a plastic fermentation bucket or glass carboy. The container is left open to breathe. Fermentation starts in a day or two and ends several weeks later. As the juice ferments, keep the container topped off with more perry. When the initial fermentation is over, rack off into another sterile container. Attach an airlock and leave the perry to mature for about six months in a cool place before bottling.

Although pears can turn into perry with nothing except the yeasts on their skins, your odds of creating a quality product increase with informed intervention, a more hands-on approach. Into every gallon of pear juice add a Campden tablet (sulfite) for sterilization. The amount of sulfite needed for perry exceeds that of cider because pears contain more of the compound acetaldehyde, which neutralizes sulfite. A teaspoon of yeast nutrient will boost fermentation. Also, a teaspoon of pectin enzyme will alleviate the haze that plagues perry. Once these additions have been made, the juice will need to stand

for another day for the Campden tablet to do its work and then dissipate so that the yeast you add will not be killed in turn.

Yeast is often "started" before being added to the juice. This involves boiling water into which a small amount of sugar is dissolved. It's important to allow the solution to cool to 105 to 110 degrees Fahrenheit before proceeding with the yeast: Too hot and the yeast will be killed, too cool and the yeast doesn't get the jump-start it needs. Consult the packet for the optimal treatment of your particular yeast. Once the water is at the correct temperature, it's time to add the yeast and the yeast nutrient (if the nutrient was not added earlier) in a fermentation bucket or other appropriate vessel. Cover the concoction and leave it alone to foam for four hours at room temperature.

This is the first of two fermentations that perry, like cider, undergoes. In this initial fermentation, the yeast converts sugars to ethanol alcohol.

The secondary fermentation comes about through a different chemical action. Now malic acid is converted to lactic acid. This happens spontaneously, without any effort on your part. First, however, you must rack off the perry to separate it from its lees (sediment). Siphon the perry into a glass carboy or other container that can be airlocked. Fill to the neck to leave as little air as possible. It's crucial that the container be sterile to circumvent certain strains of bacteria that will convert the cit-

Pear

BOTANICAL NAME: *Pyrus communis* (European pear)

PLANT TYPE: Tree

USDA ZONES: 5–9

HEIGHT: 15 feet semi-dwarf; 30–40 feet standard

SOIL: Well-drained, slightly acidic loam is best, but pears tolerate damp, heavy soil.

LIGHT: full Sun

WATER: regular

GROWTH HABIT: strongly vertical

PROPAGATE BY: grafting, seed

SPACING: 15 feet for semi-dwarfs; 25 feet for standards

YEARS TO BEARING: three to eight, depending on variety and rootstock. Fruit is produced on spurs that remain viable for about five years.

PRUNING: Train pears to a central leader. Cut to a modified leader when the tree is 8–10 feet tall. Pears require little annual pruning. Remove damaged, diseased, and rubbing branches. Remove suckers and water sprouts. Pears grow strongly upright. Vertical branches can be trained downward by attaching fishing weights until the branches reach a 45-degree angle. Pear trees can be grown as espaliers.

FRUIT THINNING: Thin when the pears are no bigger than a dime. Leave the best single fruit per cluster and one fruit every 5 inches along the branch.

HARVEST: 1–2 bushels from a semi-dwarf; 50–100 pounds or more from a mature standard. Pears do not ripen on the tree. They should be picked when a gently lifted pear snaps off at the stem joint. Store picked pears in cool, dark place, such as a basement. Within two months, bring them into a warm room three to five days before you plan to begin the fermentation.

NOTES: Pears are very long-lived trees. Two varieties are needed for pollination. Although pear trees need winter chill, their flower buds can be damaged by late frosts. If fire blight is a problem in your area, choose varieties and rootstocks resistant to the disease.

BEST USED IN: perry, pear wine, with apple in hard cider. Pear wine tends to be sweet and buttery.

ric acid of pears to acetic acid. We're not here to make pear vinegar.

After airlocking your second-stage perry, the malic acid will continue its conversion for about two weeks. The perry is then allowed to sit and clear in a cool place such as a basement or garage, often through the winter into the following spring or early summer. For centuries, perry was overwintered in outbuildings. Prolonged

Asian pears

chilling, or cycles of chilling and warming, is beneficial, allowing excess tannin to clear, but perry should never freeze. If stalled by cool temperatures, fermentation will start up again when the ambient temperature climbs to 60 degrees Fahrenheit. After a long, cold winter, you'll know your perry is done when the impurities have settled to the bottom and the liquid has lost all or nearly all of its haze.

It's now time to bottle your perry. First, sterilize the bottles. If you would like to try for a bit of bubble, add a teaspoon of sugar before racking the perry. There should still be enough live yeast to ferment the sugar and release carbon dioxide. If you do elect to effervesce, use bottles that can take the pressure and cap them tightly. If, however, you want traditional perry, skip the sparkle. Don't even think about trying for a high carbonation level, champagne-style.

The generous tannins of pears interfere with effervescence. Even worse, too much carbon dioxide can trigger lactic acid bacteria which—as we have already learned—will turn your lovely perry to vinegar.

PERRY PEARS

True perry is light, complex, and delectable. It is to pear cider what tenderloin is to cube steak. True perry does not come from the branches of a Bartlett or Bosc; it comes from perry pears. There's just no getting around it: Superior perry derives from fruit that you'd never want to bite into. The reason for this is our old friend tannin. High-tannin pears are the gold standard for perry-making. Low- to medium-tannin pears may be a fine addition to the fruit-and-cheese course, but they also make low-quality perry. Unfortunately for perry's reputation, these general-purpose pears are easier to come by and require less skill to ferment; thus, they are the basis of many commercial perries.

Perry trees can be purchased through mail-order nurseries. Or, if you're willing to turn your hand to grafting (see page 137), the National Clonal Germplasm Repository in Corvallis, Oregon, supplies scionwood of hard-to-find heritage fruit trees—including perry pears—to interested individuals. Either way, pay attention to the rootstock of your tree. Choices include standard-sized, OHxF 97 rootstocks or those from pear seedlings, either of which needs a minimum

20-foot spacing and may live more than 200 years to become grand, spreading, heirloom trees. Semi-dwarf and dwarf rootstocks are also available in the OHxF and Pyro series. Rootstocks from quince were traditionally used, but they're probably best avoided by the home grafter since not all pear varieties are compatible with quince.

Pear Varieties

The good news about perry is that, unlike apple cider, only a single variety of pear need be used. This means that one prudently picked and productive perry tree will provide what you need for an outstanding beverage. The bad news is that you will need a second pear tree planted within 40 feet for cross-pollination. But there's more good news; the second tree can be a dessert pear. Perry trees are good pollinizers. All pear varieties bloom around the same time, so your choice of spouse for your perry tree is wide open.

Perry pears tend to be small, hard, gritty, and bitter. For purposes of fermenting, they are classified as sweet, medium sharp, bittersweet, and bittersharp. Sweet pears have low acidity and are fairly low in tannin. Medium sharp pears have moderate to low acidity and fairly low tannin content. Bittersweet pears have moderate acidity and are high in tannin (there are very few pear varieties in this category). Bittersharp pears have high acidity and high tannin content. Bittersharps are con-sidered the superlative perry pear because of their well-rounded astringency. These are your best bet for single-variety perry.

In consequence of their antiquity (many of today's perry varieties grew in English orchards 300 years ago), some perry pears bear curious names. Consider the Huffcap, which will blow your hat right off your head—due to quality or potency, who knows? We can only hope that Tayton Squash is a testament to how perry is made and not a forewarning of the flavor: certainly, not even the most diehard vegan wants a heady Hubbard nose with undertones of zucchini. Perhaps Green Horse is akin to pink elephants? And Butt—let's not even speculate.

Several varieties of perry pears are available in the United States. The following cultivars are among the most popular and accessible.

GROWING PEARS

If dessert pears grow in your region, so will perry pears. Pears in general are more obliging plants than apples, although some pear varieties are less hardy than their round, red orchard-mates. Pears will grow on heavy soils in cool climates from USDA Zones 5 to 9, and some can tough it out in Zone 4. Many varieties are hardy to -25 degrees Fahrenheit, although they may resort to biennial bearing in the coldest areas. A good harvest depends on 500 to 800 hours of temperatures below 45

Pear Varieties

Barland: bittersharp. Early midseason. Ripens in late September to early October. Bears at an early age. Large, vigorous, long-lived tree. Highly scab susceptible. Fruit should be milled within three days of harvest. Perry determined as average to good quality.

Barnet: sweet. Early midseason. Ripens in late September to early October. Small fruit, easily shaken from tree. Bears at an early age. Tends to bear biennially. Large, though relatively compact tree. Scab resistant. Fruit should be milled one to three weeks after harvest. Perry determined as average quality.

Blakeney Red: medium sharp. Midseason. Ripens in late September to early October. Medium-sized fruit is crisp and juicy. Slow to bear, but reliable crops come with age. Vigorous tree. Fruit should be milled within a week from harvest. Fruit must be fully mature, but not over-ripe for perry. Perry determined as average- to low-quality, possibly due to the challenge of capturing the fruit at its optimum ripeness. It's speculated that the quality of Blakeney Red cider depends on the conditions (soil, rainfall, etc.) under which the pears were grown. The most common variety of perry pear in Britain.

Brandy: medium sharp. Midseason. Ripens early September to early October. Bears at a very early age. Heavy crops, but tends to bear biennially. Tree is vigorous, sturdy, and somewhat smaller than other perry cultivars. Moderately susceptible to scab. Fruit should be milled within four weeks of harvest. Perry determined as average quality.

Butt: bittersharp. Mid- to late-season. Ripens October. Fruit so firm it can lie on the ground for weeks without rotting. Slow to ferment. Heavy crops, but tends to bear biennially. Vigorous. Fruit should be milled in four to ten weeks from harvest. Perry determined as good quality.

Gin: medium sharp. Mid- to late-season. Ripens October. Small fruit. Large crops, but tends to bear biennially. Stores well. Vigorous, medium-sized tree. Scab resistant. Fruit should be milled three to five weeks after harvest. Perry determined as good quality.

Green Horse: medium sharp. Mid- to late-season. Ripens October. Consistent crops. Stores for several weeks after harvest. Grows to a large tree. Scab susceptible. Fruit should be milled within three weeks of harvest. Perry determined as good quality.

Hendre Huffcap: sweet. Midseason. Ripens late September to October. Small, elliptical fruit is easily shaken from the tree. Vigorous, large tree. Resistant to scab. Although low in tannins, this variety still makes a good, single-varietal perry. Fruit should be milled within two weeks of harvest.

Tayton Squash: medium sharp. Early season. Ripens September. Small fruit does not store well. Bears early and productively, but usually biennially. Fruit must be picked as soon as it's ready as it squashes upon falling, thus the name. Medium to large tree. Very susceptible to scab. Fruit contains a high level of citric acid. Considered to make very fine perry: sweet, clear, and strong. Fruit must be milled within two days of harvest.

Thorn: medium sharp. Early to midseason. Ripens September. Its small fruit can also be used for cooking (though not optimal). Small, compact tree bears heavily, but somewhat later in life than many other perry pears. Susceptible to scab. Perry determined as good or very good quality. Fruit should be milled within a week of harvest.

Winnals Longdon: medium sharp. Midseason. Ripens early October. Small fruit. Medium to large tree is productive, but tends to bear biennially. Slow to reach fruiting age. Scab resistant. Perry determined to be good quality. Fruit should be milled within a week of harvest.

Yellow Huffcap: medium sharp. Midseason. Ripens late September to early October. When ripe, the fruit must be picked or shaken from the tree or it will rot where it hangs. Vigorous, the large tree is productive, but a biennial bearer. Slow to reach fruiting age. Small, elliptical fruit contain a high level of citric acid. Considered to make very fine perry with a kick. Fruit should be milled within a week of harvest.

Golden russet pears

degrees Fahrenheit. Pears also require the right balance of sun and rain. They prefer cool, moist, cloudy weather. Perry pears have a memory: the quality of the summer weather will be the heart of the perry that will follow.

Pears bear at a slightly later age than apples. Once they reach puberty, they bloom up to three weeks earlier than apples, making the buds and flowers susceptible to late frosts. On the plus side, pears require less pruning and spraying than apples, and perform better on poorly drained soils and clay. Ideally, plant your pear tree in deep loam, but they will soldier on in fast-draining, sandy, and acidic soils. If, however, your site is downright boggy, abandon all hope of growing a pear tree. Barland, in particular, resents waterlogged soil. Most pears are extremely susceptible to fire blight,

a devastating disease that lays waste to almost any part of the tree. If you live in an area plagued by fire blight, plant only resistant varieties and rootstocks.

As the saying goes, pears are planted for heirs. This is a reminder to choose the site for your perry tree wisely, as it may provide crops not just for you, but for your children, grandchildren, great-grandchildren, and generations to come. After 100 to 200 years or more, perry trees become venerable, canopied giants that lavish the neighborhood with a ton of pears each year. No, really—a literal ton.

If you don't have the room—or ancestral imagination—for a legacy tree, there are also semi-dwarfs. Perry pears on semi-dwarf rootstocks mature to about 15 feet, and can be set 10 to 15 feet apart. They produce fruit about a year earlier than their standard-sized kin. Within a few years, a semi-dwarf should be gifting you with a bushel or two of pears.

Choose a planting site with at least eight hours of sun a day and good airflow. Low areas can be subject to late frosts, which can blast early pear blossoms. Dig a hole the depth of the rootball but wider. Set the rootball in the hole with the top (or top of the roots, in the case of a bareroot sapling) level with the ground surface. Don't plant too deeply and be sure the graft union is above the soil. Spread out the roots. If the roots have begun to encircle the pot, work them loose as gently as possible and head them away from the rootball.

It's best to avoid staking a tree, which encourages it to rely on the support rather than its own root growth. However, if it appears your new tree may keel over from insufficient anchoring, this is the time to put in the stake—when you can see where the roots lie. Fill the hole halfway with soil and then add water all the way to the top; this will ensure the soil is thoroughly settled around the roots and no air pockets remain. Once the water has cleared, fill the hole with the remaining soil. Make a bowl-shaped depression from the tree trunk to the perimeter of the rootball to make watering easier. You'll need to keep the soil somewhat moist through your tree's first summer, if not several subsequent summers. Give your new pear tree a little extra love in the form of a 2-inch layer of mulch spread evenly over the planting area. Manure or compost are best, but wood chips, shredded leaves, straw, or lawn clippings will do.

For the first few years, train your pear to a central leader. This simply means letting one centrally placed shoot grow upward, while removing or shortening any other shoots that attempt to outgrow the leader. Once your tree is 8 to 10 feet tall, cut the leader back to a strong and highly placed lateral shoot growing from it. You now have a modified-leader pear tree.

Pears require little pruning. Follow the usual rule of removing anything damaged, diseased, or rubbing. Suckers coming from the trunk or the rootball should be rubbed or pulled off as soon as they appear. Pears don't throw a lot of water sprouts (vigorous, vertical stems growing upward from the center of a branch or the trunk), but when one appears cut it off close to the branch, trim it back to a well-placed bud, or bend it over into a useful, horizontal new branch.

There's one other thing to know about training your tree. Young pears stand to attention. That's to say, they grow strongly upright. While you don't want to dampen their spirit, you must consider their future. The scaffolding you create today will be the structure of your pear for the next 250 years. Talk about pressure. Actually, it's an easy fix. Fishing weights and Christmas ornament hangers will do the job nicely. Simply weigh down the sky-reaching branches to a 45-degree angle. A 60-degree angle is optimal, but the weights will usually pull the branches down a bit farther as the branch relaxes. Leave the weight on for about a year, by which time the adjustment should be permanent. Once the tree begins to bear, the weight of the fruit will take over the job.

Now that your tree is studded with embryo fruit, it's time to thin that fruit. Thinning encourages larger, better quality pears and prevents the branches from breaking under heavy loads. It can also stem the tendency of some pear varieties to biennial bearing. Thin early when the pears are no bigger than a dime. Leave the best single fruit per cluster and one fruit every 5 inches along the branch.

CHAPTER 7

Liqueurs and Infused Spirits
From Garden to Glass

Liqueurs, often thought to be fancy and complicated, are some of the simplest drinks to make. Unlike beer, cider, wine, and perry, liqueurs are created with ready-made alcohol. All you need to do is add a cup of this or a dash of that. Then pour on the alcohol, seal, and wait a while. Once it tastes the way you like, add sugar for liqueur (or leave it out for an infused spirit), and you're done!

Liqueurs can go with just about everything: sophisticated dining, as a dessert, in a dessert, poured over ice, mixed with coffee or espresso, or poured with mixers and blended with other liquors for cocktails. Share your best concoctions and you will look like a culinary genius to your friends and family. You'll find the process itself—turning liquor into liqueur—is a creative outlet and a culinary art form. Once you get the hang of it, chances are if you can grow and eat it, you can turn it into a liqueur. You will wonder why you never made it before!

The best part about growing your own ingredients for liqueurs is that it can come straight out of your garden and into a bottle. Nothing is fresher than only a few footsteps between harvest and glass! Since herbal liqueurs are the easiest to make, and coincidentally the easiest to grow, we focus on all sorts of herbs in this chapter. However, there's a wild world of recipes out there. If you find something promising that uses fruit—or even hops—scan the other chapters or the index of this book for information on the plants you might want to grow.

HISTORICAL LIQUEUR

Liqueurs, sometimes referred to as cordials, are comprised of three key components—liquor, flavoring, and sugar. Infused spirits use many of the same ingredients; however, without the sweetener. Whether you realized it or not, you've likely sampled one or more liqueurs already. Different regions of the world have their own signature liqueurs: O Cha (Japanese green tea liqueur), Forbidden Fruit (American apple and citrus-flavored cordial), Jägermeister (gentian-flavored German liqueur), and

Kahlua (a strong coffee-based Mexican liqueur) are just a few of the commercially produced liqueurs from around the world.

Liqueur, not to be confused with liquor, is a French word derived from the Latin word *liquifacere*, which means to liquefy—melt or dissolve. However, the French weren't the first to make liqueurs or spirits. The Chinese had distilled them as early as 800 BC, and grape wine was distilled by both Greeks and Egyptians.

Most likely, the first liqueurs came from Greece and were made from distilled wine, honey, and fruit. The ancient Greek physician Hippocrates (460 to 370 BC) added cinnamon to fermented honey, and to this day, mead mixed with fruit and spices is called a Hippocras.

In Europe, liqueurs were considered medicinal and in the earliest references to distillation, monks and alchemists produced the spirits and the distillation was similar to what we would call moonshine—and with a nasty taste. Flavoring the medicinal spirits came about to circumvent those horrid bitter tastes. When trade opened up to India, all manners of aromatic spices and sweeteners like sugar cane provided flavorings that helped the medicine go down.

From there, some liqueurs started making a name for themselves. One of the world's best-known liqueurs, Bénédictine, was produced by a monastery in Fécamp in Normandy as early as 1510. Their trademark, D.O.M., stands for *Deo Optimo Maximo*. This translates as: To God, Most Good, Most Great. Made from a fine cognac, this is the only ingredient revealed; it is unfortunate that the other components are closely guarded secrets—yet that is befitting an elixir to a god. We mortals are fortunate when we can enjoy this liqueur!

Chartreuse is another liqueur originally developed for medicinal purposes. It was made by Carthusian monks in the seventeenth century and to this day has a closely guarded complex recipe that contains 130 different herbs. (Oh, and that yellow-green color called chartreuse is actually named because of its resemblance to the hue of this original drink developed by the monastery la Grande-Chartreuse.) By the nineteenth century, the monks had turned Chartreuse into a commercial enterprise. These days, the monks distill the liqueur and have an outside company bottle, distribute, and sell it as Les Pères Chartreux. The profits from the drink help finance charitable work.

BRINGING LIQUEUR TO LIFE

We probably won't be growing 130 different herbs in our home gardens, but Chartreuse liqueur is proof that the ingredients we can grow and try with our own liqueurs are endless. Experimenting with mixtures of compatible herbs or fruits, we can spend a lifetime mixing our own exotic flavors grown from our very own gardens.

Making most liqueurs is very simple as well: You put the ingredients together, let them sit for a week, strain the liquid, bottle, and seal. For example, to make Praise of Fraise, you gather 2 pints of fresh strawberries straight from your garden and combine them with 1½ cups of simple syrup and one-fifth of vodka (at 80 proof or more). In a ½-gallon jar, muddle the strawberries and syrup with a muddler (a pestle-like tool) or use a spoon to mash. Seal the jar with a lid and place in a cool dark place for approximately seven days. Test after about a week to make sure it has a strong taste and scent of strawberries. Then strain the liquid with a mesh strainer, being careful not to touch the solids with your hands. If you seal it in a sterile bottle, it can keep in a dark cabinet for up to one year. Perfect for a pop of summer in the colder months!

For most of the herbs that follow, we'll provide at least one guideline recipe, but

Simple Syrup Recipe

Since granulated sugar does not completely dissolve in alcohol, you can dissolve it in water and make a simple syrup. Mix 2¼ cups granulated cane sugar with 2¼ cups water. Bring the mixture to a boil and stir until sugar is dissolved. Let the liquid cool and refrigerate for up to three months.

it's all about your own tastes, and how pungent and fresh your own ingredients are. Play with ratios of fresh herbs, vodka, and sugar (or simple syrup) until you find what works for you. Feel free to scale down and perform smaller experiments as well!

Which Liquor Should You Use?

When making liqueur or infused spirits, you should choose your base liquor based on what you are trying to accomplish in taste. A neutral spirit such as vodka or white brandy in the recipe allows the other flavorings to take center stage. However, you can enhance or complement your ingredients with brandy, whiskey, rum, or tequila. As long as you choose an 80 proof or more

alcohol (forty percent alcohol content), it will be strong enough to act as a preservative. (And it may seem counterintuitive, but with vodka the higher proof versions often will create a smoother liqueur.)

Since this book focuses on the growing side—we are gardeners and not distillers, after all—we recommend you buy or borrow a good liqueur recipe book if you're interested in infusions beyond the basics. Our favorite recipes come from innovator Andrew Schloss: *Homemade Liqueurs and Infused Spirits*. With 159 flavors to create, you could spend a lifetime creating liqueurs to dazzle your taste buds. You'll find most recipes are not complicated, and take a minimum amount of equipment.

ON TO THE HERBS

The herb garden is going to be one of the most useful for making liqueurs. For most herbs, growing them is as easy as making liqueur! Some are rampant growers; others behave themselves and thrive on neglect. Starting an herb garden is easy enough. Most herbs you grow need well-drained, humus-rich soil. Except for a few herbs, save adding fertilizers for your vegetables; the herbs will like a compost-rich soil instead! Grow your own herbs in containers or tucked in with your ornamentals or vegetables. These plants are great for beginner gardeners, but seasoned gardeners grow them too.

Soil pH level isn't as much of a worry if you are regularly adding compost to the beds, which tends to neutralize the pH balance. Rainy areas typically will have acidic soil as the rain leaches out the calcium. On the flip side, arid regions tend to have alkaline soil. In looking for the Goldilocks level of just right, most herbs prefer the pH level to be between 5.5 and 6.5. However, in new herb beds where compost hasn't had time to work its magic, you will want to test your soil pH level. If too acidic you can add dolomite limestone, and too alkaline you can add sulfur until your annual applications of compost helps change the soil and neutralizes the pH levels. Adding acidic mulches to the top of the soil such as pine needles will help. Alkaline soil is toughest to keep at a lower pH level though,

as the limestone is constantly dissolving and sweetening the soil.

Grow Herbs in Pots

With rampant growers such as mint that send out underground shoots and spread in an effort to take over the world, you will want to tame them by growing them in deep containers that don't come in contact with soil. One plant can grow and quickly fill a pot in one season. Better the container than your garden. After a year or two, depending on the width of the pot, the plants become too congested and will crowd themselves into oblivion. When the plants start to decline, you will know it's time to divide them. Lift them out and divide out a few plants for your new planting. Unless you have a hot compost pile that will kill the roots of rampant growers like mint, discard the soil, roots, and plants, or give starts to friends. Be sure to warn them about how rampant the plants are, or a friendship could be lost! When repotting your divisions, use fresh potting soil. You can mix old and new potting soil in the bottom of the pot to save on costs and save the new potting soil for where most of the roots grow.

When growing any herbs that produce better flavor if grown with little to no fertilizers, don't use a potting soil with timed-released fertilizer. Fresh potting soil will be fertile enough to get your young plants growing when first planted and you can add a conservative amount of organic fertilizer to the containers early in the growing season. Herbs grown in rich soil will not produce as much volatile oils that contribute to plant's flavor.

As with any gardening endeavor, start small, build on your successes, and learn from your failures.

Harvesting Herbs

For many herbs, the best time to harvest is during the early blooming stage, when the volatile oils are at their peak. After the flowers begin to develop seeds, the flavor of the leaves deteriorates. When the bottom-most leaves at the base of the plant begin to fade and fall from the stems, the plant has reached optimum development. Cut your stems with garden shears in the morning when the dew has dried, when the essential oils are at their peak. Take care not to bruise the leaves. Harvesting in the morning means less wilting and less bruising of the leaves. If the weather is cool, you can wait until later in the day before harvesting.

Drying Herbs

You won't always be able to use fresh herbs in your tinctures. Plus, you may want to make some batches during the off season months.

To dry herbs in small quantities, wash them and pat them dry, tie them together in small bundles, and hang them to dry upside down in a dry, dark place. Light will fade the leaves if exposed to sunlight. You can also spread your herbs out on drying screens. Drying can take up to fourteen days to complete. Fans help move the air around in rooms and will help the drying process. Don't let the fans blow directly on the product as you might end up with it blown off the screens. High humidity areas will need good airflow over the leaves. To complete the drying process the humidity needs to be less than forty percent. Drier air speeds up the process. The faster the herb can dry the better quality it will be.

Remember dried herbs have a higher concentration of flavor than fresh. Use fresh whenever you can, and adjust your measurements in recipes accordingly for dry or fresh.

Angelica

With a name like Angelica, this herb must make a drink fit for the angels! Perhaps that's why many liqueur recipes feature Angelica as one of their ingredients. For example, the herb is reputed to be the main herb in the secret recipe for Chartreuse liqueur.

Although this tall, edible herb species is ornamental with bamboo-like stems and umbels filled with small white flowers, be sure you are growing *Angelica archangelica* and not an ornamental species such as *A. pachycarpa* or *A. gigas*. All parts of the sculptural, 6-foot-tall, biennial herb are edible—the roots, seeds, and stems. Related to the carrot, Angelica forms a tap root—and this is the part most commonly used to make liqueurs and infused spirits. The amounts used will depend on your recipes and what part of the plant is used; recipes can call for as little as 2 tablespoons and up to ½ cup of its fresh roots.

Being a biennial plant means the herb will grow its foliage the first year and set seed and die the second year. If you don't allow it to go to seed, the plant will live another year.

It's best to grow Angelica in your ornamental, herb, or kitchen garden. The corymb of white flowers attracts pollinators, so it is useful if grown close to other plants that need pollination to bear fruit or seed. It is not advisable to harvest this plant in the wild. Since the herb closely resembles the poisonous water hemlock (*Cicula maculata*), you wouldn't want to mistake its identity with the other and wind up with deadly results!

Angelica

BOTANICAL NAME: *Angelica archangelica*
PLANT TYPE: biennial herb
USDA ZONES: 4–9
HEIGHT: 6 feet
SOIL: moist, rich, slightly acidic loam. Side-dress plants with a heavy layer of compost to retain moisture.
LIGHT: part shade, full sun in cool climates
WATER: regular to heavy. Unlike other herbs that prefer drier soil, angelica needs weekly irrigation.
GROWTH HABIT: upright, leafy
PROPAGATE BY: fresh seed, root divisions, and cuttings in early spring. Self-seeds readily. Most seed will not germinate until the following spring.
SPACING: 18 inches

MONTHS TO HARVEST: eighteen
PRUNING: none needed. The lifespan of the plant can be extended to a third year if the flowers that appear the second year are removed.
HARVEST: The stems and leaves should be collected before the plant flowers.
NOTES: Angelica prefers regions with mild summers. Foot-wide, roundish umbels composed of tiny white flowers make the plant ornamental. The easiest way to get started growing this tall herb is to purchase young plants the first two years and let them flower and go to seed. Allow some seeds to fall to the ground and sprout.
BEST USED IN: liqueurs

Freezing Herbs

Fresh herbs give the best flavoring, but a trick to keeping the freshest taste in storage is to wash your harvest, chop the herbs, measure, and add to water-filled ice cube trays. When the herb cubes are solidly frozen, remove from ice tray and place in freezer storage bags or containers. When ready to infuse your liqueurs, thaw out the number of ice cubes you wish to use for your batch, strain the water, and add the herbs to your liqueur. Herbs, such as basil, that lose flavor when dried will keep a fresh taste when frozen. To freeze, rinse your freshly picked basil or other herbs, and remove tough stems. Spin dry using a salad spinner, or pat dry between clean towels or paper towels. Chop basil into fine pieces and drop up to 2 teaspoons of chopped basil into the water in each cube in the tray. Keep your measurements uniform throughout the whole tray. Once frozen, package the basil ice cubes in your air-tight freezer storage container and mark on it how much basil is in each cube. You can use this method for almost any herb you grow and want to keep its fresh taste long after harvest is over.

Anise

Ho, ho, ho and a bottle of Pimpinella-infused rum? Yes, anise can spice up a bottle of rum! It is also used in a national drink made in Greece—Ouzo. In fact, in ancient times the Greeks used licorice-flavored anise in bread and wine. Now it is used in many dishes to help sweeten them, or for its licorice-like flavor in the case of liqueurs. Anisette is a French liqueur flavored with anise seed.

Anise is an annual herb whose flower clusters look similar to Queen Anne's Lace to which it is related, and has naturalized like a wildflower across the country. And like Queen Anne's Lace (*Daucus carota*), once planted in your garden, anise plants will likely seed themselves every year if you don't harvest them and let them go to seed. Pull out any unwanted ones. They won't transplant because of their carrot-like tap root.

Once widely grown commercially, anise is often replaced commercially by cheaper anise-like flavoring such as *Illicium verum* and synthetic flavorings. When you grow it, you are using the authentic herb, not some cheap knockoff. That doesn't mean you can't make a great knockoff drink, however: This Ouzo liqueur knockoff has plenty of licorice flavor. Crush ½-cup licorice root and ¼ cup anise seed and crack twelve star anise seeds. Place in jar and pour one-fifth vodka over it. Seal the jar and place in a cool, dark place. After three to five days the infused vodka will smell and taste like licorice. Strain the liquid and stir in 1 cup simple syrup. Bottle and store the liqueur in a dark, cool place for up to one year.

Anise

BOTANICAL NAME: *Pimpinella anisum*

PLANT TYPE: annual herb

USDA ZONES: 7–10

HEIGHT: 18–24 inches

SOIL: light soil

LIGHT: full sun

WATER: regular. The germinating seed needs constant water; established plants can tolerate low water or short periods of drought.

GROWTH HABIT: feathery and wispy

PROPAGATE BY: seed, planted when the soil is at least 60 degrees Fahrenheit

SPACING: 1 foot

MONTH TO HARVEST: four

PRUNING: none

HARVEST: In August or September the dry flower heads can be placed in a paper bag. When thoroughly dry, shake out the seed.

NOTES: The fruit of anise (often incorrectly referred to as seeds) adds a licorice flavor wherever it's added. In fact, anise oil is the predominant flavoring of black licorice candy. Anise may be grown in pots and moved outdoors in late spring.

BEST USED IN: liqueurs, beer

Basil

Basil is a heat-loving herb in the garden, and quite easy to grow. It makes a good container plant as well.

It's best to plant basil seeds once the weather has warmed at the end of spring as germination is poor to nonexistent in cool soil. For an earlier harvest, plant seeds indoors in pots two to three weeks before the last average date of frost for your area. Cover the seeds with ⅛ inch of seed-starting mix or sprinkle seeds over the top of the mix. Then you're ready to plant outside as soon as the soil has warmed to at least 50 degrees Fahrenheit!

As far as taste, fresh-picked sweet basil has a wonderful taste that's hard to describe. It's a sweet, spicy, licorice, clove kind of flavor. So sure, basil is what lends flavor to many savory dishes, like tomato and pesto sauces. Yet it turns out that basil can be equally tasty combined with fruits like cherry in a liqueur. These two powerful flavors can make a fantastic combination. Don't believe me? Try it yourself: Combine one bunch fresh basil, two pints of sour

cherries stemmed and crushed (leaving pits in), and place it in a jar with one-fifth of 80- to 100-proof vodka. Infuse for up to five days, until you can taste and smell the cherries and basil. When the infusion is finished, strain the liquid and add 1¼ cup simple syrup—or more to taste.

Basil is used as an ingredient in quite a few other liqueurs as well. We've seen it married with strawberries and lemons, and in a specialized basil liqueur made with Genovese basil leaves.

Basil loses too much flavor when dried, so fresh is best, and frozen the next best thing.

Basil

BOTANICAL NAME: *Ocimum* spp

PLANT TYPE: tender perennial herb, often grown as an annual in cooler zones. Perennial in Zone 10

USDA ZONES: 4–10

HEIGHT: 12–24 inches

SOIL: moist loam

LIGHT: full sun

WATER: regular to high

GROWTH HABIT: low, leafy

PROPAGATE BY: seed

SPACING: 6–10 inches

MONTHS TO HARVEST: two to three months from seed

PRUNING: Harvest by pinching to keep plants bushy.

HARVEST: Pinch leaves and stems as needed. When the plants begin to flower, they cease leaf production. Prevent basil from blooming for as long as possible by harvesting or pinching off the top sets of leaves as soon as the plant reaches about 6 inches in height. Freeze basil leaves for later use rather than drying them, which changes the flavor.

NOTES: To encourage bushiness, pinch out the tip of the seedlings once they reach 6 to 8 inches.

BEST USED IN: liqueurs

Caraway

This old world biennial herb, used in the Middle East for thousands of years before it was introduced to Europe, was named after the ancient region Caria that is now Turkey. In its native haunts, caraway sets its roots down in moist grasslands and disturbed soil. Oh, for the old days when potions made with caraway attracted love! All's not lost though; we can still use the seed to flavor dishes and make a wonderful Dutch-style liqueur.

The word *kümmel* means caraway and cumin seed in Dutch, German, and Yiddish languages, and they are the two of the main ingredients in Kümmel, the original caraway liqueur. To make a knockoff of the original takes only hours to tinc-

ture as the caraway is so strong. Add ½ cup each of the caraway seed, cumin seed, and fennel seed, one-fifth of vodka (80 to 100 proof) and ¼ cup simple syrup. Four to eight hours later, this liqueur is ready to serve! Adjust to taste with additional simple syrup and serve, or store your Kümmel in a sealed bottle for up to one year.

Caraway

BOTANICAL NAME: *Carum carvi*

PLANT TYPE: biennial herb (grows the first year; flowers the second, after which it sets seed and dies)

USDA ZONES: 3–8

HEIGHT: to 8 inches the first year, 24–36 inches the second

SOIL: well-drained, rich loam

LIGHT: full sun

WATER: low to moderate

GROWTH HABIT: upright, leafy

PROPAGATE BY: seed, best sown in autumn. Cuttings.

SPACING: 12 to 18 inches. Sow several seeds per spot and thin to strongest seedling.

YEARS TO HARVEST: two

PRUNING: none

HARVEST: Collect the seeds when ripe, but before they fall to the ground. Remove the flower heads and gently shake to catch the seeds. Dry-roast the seeds to enhance their flavor.

NOTES: Caraway "seed" is actually the plant's fruit.

BEST USED IN: liqueurs

Cilantro growing next to raspberries

Cilantro/Coriander

Cilantro is one of those tastes you either love or seriously dislike. Our genetic makeup can mean the difference between tasting a wonderful herb or tasting what seems like a bar of soap. If you are one with the genes that make you gag over the taste of cilantro, it's probably smart to pass on growing this or adding it to your liqueurs!

All parts of the herb are edible and both the seed and the leaves are used in liqueurs. The seed is called coriander, and the leaves are referred to as cilantro and they have different flavors. We prefer to use fresh leaves, but freezing the leaves for later use is another option. Cilantro and lime are good complementary flavors for each other, which is why you'll find plenty of recipes that call for the two. Start experimenting with the herb once you've tried some recipes, and you will find it makes for an interesting complex-flavored liqueur. How about some tequila? One-fifth of tequila (80 proof and no worm), ½ cup toasted and ground coriander seeds, a chopped bunch of fresh cilantro, grated zest of 3 limes, and 1 cup simple syrup. Pour the tequila over the seeds, cilantro, sea salt, and lime zest in a jar and seal. Leave in a cool, dark place for three to five days. It should have a strong coriander taste when done. Strain liquid and stir in the simple syrup. Bottle and seal the liqueur and keep in a cool, dark place where it will keep up to one year.

Cilantro

BOTANICAL NAME: *Coriandrum sativum*
PLANT TYPE: fast-growing annual herb
USDA ZONES: all
HEIGHT: 12–18 inches
SOIL: rich, well-drained but moist
LIGHT: full sun or light shade in the hottest areas
WATER: regular
GROWTH HABIT: erect or bushy
PROPAGATE BY: direct-sown seed. In some regions, cilantro will self-seed.
SPACING: 6–8 inches
MONTHS TO HARVEST: Harvest leaves in about three to four weeks; seed (coriander) in about forty-five days.
PRUNING: Pinch back young cilantro plants an inch or so to encourage fullness. Unless you're growing the plant for coriander seed, snip off the top of the main stem as soon as flowers begin to appear to redirect the plant's energy into leaf production.
HARVEST: When the plants are large enough, pinch back and pick off leaves as needed. Use the new, finely cut leaves in cooking, rather than the ferny lower leaves. Cilantro loses most of its flavor when dried. For coriander seed, harvest on a dry day. When the seedpods begin to brown, cut off and crack to release seed. Allow the coriander to dry in a cool, well-ventilated place.
NOTES: Cilantro is the common name of the leaves of *Coriandrum sativum*; coriander is the common name of the seeds of the same plant. Cilantro bolts (goes to seed) quickly in hot weather. It resents being transplanted; however, an early harvest can be achieved by planting seed in peat pots about two weeks before the average date of last frost. For a steady supply of fresh leaves, make successive sowings of cilantro seed every two to three weeks from spring to fall. Fall sowings will over-winter in warmer areas and begin growth the following spring.
BEST USED IN: liqueurs

Lemon verbena

This shrub's unmistakable lemony fragrance makes itself known anytime you brush against it. It's such a pleasing aroma, that it's no surprise that dried lemon verbena leaves are often used in potpourris and to flavor fruit-based drinks, teas, and liqueurs as well.

Spaniards, who introduced the plant to Europe, discovered these South American plants native to Chile and Argentina. The herb was named after the Princess Louisa of Parma, the wife of King Carlos IV of Spain. The botanical name *Aloysia* is the Latin equivalent for the princess's first name, and to this day Europeans commonly call the shrub herb Louisa. During the Victorian era, lemon verbena was simply known as the lemon plant. It now has plenty of common names: Cedrón, herb Louisa, hierba Luisa, lemon-scented verbena, Louisa; and past names: *Aloysia citriodora*, *Lippia citriodora*, *L. triphylla* *Verbena citriodora*, and *V. triphylla*.

Lemon verbena

Lemon verbena

BOTANICAL NAME: *Aloysia tryphylla*

PLANT TYPE: tender, shrubby, deciduous herb

USDA ZONES: hardy perennial in Zones 9 and 10. Can be grown as a summer annual or houseplant in colder zones.

HEIGHT: 4–8 feet

SOIL: well-drained

LIGHT: full sun to light shade in hottest areas

WATER: regular during growing season

GROWTH HABIT: sparse, shrubby

PROPAGATE BY: semi-hardwood cuttings taken in midsummer

SPACING: 5 feet

MONTHS TO HARVEST: two to three

PRUNING: These naturally leggy plants need regular shearing to keep bushy. Throughout the growing season, snip stems back to keep the plant in shape. Conversely, it can be pruned as a standard.

HARVEST: Snip off leaves as needed.

NOTES: Lemon verbena has a strong lemony scent and can be a great substitute for lemons. It can be grown in a pot and overwintered indoors or treated as an annual and replaced every spring.

BEST USED IN: liqueurs

For your own liqueurs and infused spirits, use freshly picked leaves when available, or dry in small bundles or on drying racks for infusion after the growing season ends. Fresh leaves can also be stored in sealed bags or containers and refrigerated for a short time if you can't use them right away. You can harvest the leaves anytime during the growing season, but the strongest flavor and scent in the leaves comes when the plant is in bloom.

An especially refreshing simple recipe for a liqueur calls for ½ cup lemon verbena leaves, 4 cups vodka, and 2 cups sugar. Combine and seal in a sterile jar for two weeks, occasionally shaking to help dissolve the sugar. When it's ready, the infusion tastes like the citrusy fragrance of the leaves. Drain the liquid and bottle it in airtight bottles or jars. For an even more flavorful infusion that may be ready in as little as a week, add one 4-inch strip of lemon peel (without any pith) to the infusion as well. Sealed in a sterile jar or bottle and stored in a dark cabinet, the liqueurs will keep for up to one year.

Licorice

If you have a sweet tooth, here's an herb made just for you. The root stems are fifty times sweeter than sugar! What is surprising is that this herb is a member of the legume family—related to peas and beans. However, this plant is not an annual. You typically grow the plant for three years before you begin harvesting the rhizomes. While growing, keep the flowers cut off to encourage the plant to send its energy into root production, not flower production.

You can use fresh roots, but you will want to dry them for later use throughout the year. The dried root pieces will keep for one year. The herb sends its roots 3 feet down and the rhizomes can wander up to 30 feet away, with little licorice plants cropping up far away from the mother plant. The plants won't completely take over the garden, but it can be annoying.

Use a shovel when harvesting your plants; a fork will not work for digging them up, as the roots are too long and deep in the soil. This is when you will be glad you have good, loose soil, as you need to dig deep. In the fall, dig a deep trench around the plant, then dig following the roots as

Licorice

BOTANICAL NAME: *Glycyrrhiza glabra*
PLANT TYPE: deciduous herbaceous herb
USDA ZONES: 7–10
HEIGHT: 2–6 feet
SOIL: deep, fertile, well-drained. Licorice performs poorly in clay soil.
LIGHT: full sun; tolerates part shade
WATER: moderate to high
GROWTH HABIT: erect, bushy
PROPAGATE BY: seed sown in late summer to early fall in spring; rhizomes; division
SPACING: 18 inches
YEARS TO HARVEST: two to three. The roots become tough and fibrous from the fourth year on.

PRUNING: none
HARVEST: Licorice root is harvested in the autumn.
NOTES: Licorice root can be used fresh, but is typically dried. The dried roots hold their quality for a year. Pregnant women should avoid licorice. Other culinary licorices include Russian licorice (*Glycyrrhiza echinata*), American licorice (*G. lepidota*), and Chinese licorice (*G. uralensis*). Any licorice in the garden can be difficult to eradicate once established.
BEST USED IN: liqueurs, beer

they span out away from the plants and harvest as you go. You can select and set aside some rhizomes for replanting. When you are done, your trench that will look like something you've dug in for a war. Fill your soil back in.

Before drying the roots, you will want to cut them into manageable pieces for your liqueurs. After the root dries it will be a tough, fibrous material that is difficult to cut through. You will want to dry them using a heat source that is close to 95 degrees Fahrenheit but not higher.

The plant's botanical name comes from the Greek words *glykys* for *sweet* and *rhiza* meaning *root*. The Greeks and Romans loved the plant for its sweetness. It's not surprising that this herb found its way into liqueurs. It doesn't take much licorice root to make a licorice liqueur: Just 2½ tablespoons of washed and finely chopped licorice roots. Pour 1½ cups vodka over the roots and let sit in a sealed jar for one week. Strain the liquid and add ½ cup simple syrup to the spirits and let it sit for another week. Bottle it and store in a dark, cool place for up to one year.

Mint

Mint is a plant that wants to romp around your garden. Left to its own devices in the ground, it will invade your lawn, trample other plants, and if it loves the ground it grows on, will colonize it completely. We love our mint flavorings though, and so we grow it and imprison it in deep containers that don't have any contact between the drain hole in the bottom and the ground—where the mint might try to sneak out in search of new territory.

Either use the mint leaves fresh, freeze in ice cubes, or dry by harvesting stems and tying them in small bundles to dry if you plan to use them later in the season. (Though fresh is always best for liqueur in our opinion!) Buy your plants from a reliable source and from there all you need to decide is how much mint flavor you want in your liqueurs.

As far as liqueurs focused on mint, crème de menthe is the most famous one and is easy to make. Put 2 cups of fresh mint leaves into a jar and cover with 4 cups vodka (80 to 100 proof). Let it infuse for two days. Strain liquid and add 1½ cups simple syrup, or more if you want it sweeter. If you want a festive green liquid, add drops of green food color until you get the color you want.

Mint

BOTANICAL NAME: *Mentha* spp.

PLANT TYPE: perennial herb

USDA ZONES: Peppermint is hardy to Zone 3. Spearmint is the best mint for heat, up to Zone 11.

HEIGHT: 1–3 feet depending on species

SOIL: will grow in almost any soil, but prefers moist, rich loam

LIGHT: full sun to shade

WATER: regular. Mint prefers more water than do many herbs.

GROWTH HABIT: spreading

PROPAGATE BY: division, runners, cuttings, seed

SPACING: 18–24 inches

MONTHS TO BEARING: two

PRUNING: Harvesting serves as pruning. Cut old stems to the ground in winter before new growth begins. If not harvested, shearing back by half in June or July and keeping watered and fed will bring a fresh flush of new foliage.

HARVEST: Snip stems as needed. Choose fresh, unblemished leaves.

NOTES: spreads rapidly by underground stems. Best contained in a pot. Choose from spearmint, peppermint, apple mint, orange mint, and others. Peppermint is considered to be the best for classic mint flavor.

BEST USED IN: liqueurs, infusions

Rosemary

In warmer parts of the country, rosemary is a beautiful evergreen. The shrub-like perennial grows very well in many West Coast gardens. The flower hues range from dark to light blue to white depending on the variety. The plant takes well to container culture too, so climate-challenged gardeners can grow the herb in a 12-inch diameter pot outdoors during the growing season. To protect it from the ravages of cold winters, the plant is moved into a greenhouse or a sunny window indoors for winter.

Because rosemary likes a bit of humidity and heated indoor air is dry, it helps the plant to set the container on a tray of pebbles with water below the surface to keep the humidity levels up around the plant. Don't let the container dry out completely any time of the year. After a few years the rosemary grows lanky and woody. When that happens, most people replace the plants with new ones as they don't like the looks of them. You can train and prune the plants into fanciful topiaries or let them grow naturally for easier harvesting. Rosemary dries easily by tying bundles of sprigs of young needles together and hanging upside down.

Rosemary has a strong, piney scent that can overpower many flavors. There are many recipes that pair rosemary with lemon or orange citrus hoping that they embrace each other rather than one overpower the other—or turn into a pine-smelling cleaner!

Use the younger smaller leaves for your infusions, cutting fresh sprigs about 4 inches long and use sparingly so as not to end with an overpowering resinous taste. The amount you use will vary between recipes. Vodka, lavender, and rosemary make up this lavender- and rosemary-infused vodka, which is simple to make. Pour one-fifth of vodka (80 to 100 proof) over 1 sprig of rosemary, and 2 sprigs of lavender. Seal jar and infuse in a dark, cool place for three to five days. When you can taste the rosemary and lavender, strain the liquid, bottle and seal and keep in cool, dark place where it will keep up to one year.

Rosemary

BOTANICAL NAME: *Rosmarinus officinalis*

PLANT TYPE: evergreen woody herb

USDA ZONES: 7–10. Zone 6 in a sheltered microclimate.

HEIGHT: 1–8 feet

SOIL: sandy, well-draining

LIGHT: full sun

WATER: low to moderate

GROWTH HABIT: shrubby

PROPAGATE BY: seed, cuttings, layering (plants will self-layer where branches touch the ground)

SPACING: 2 feet or more, depending on variety

MONTHS TO HARVEST: four to five

PRUNING: Pinch tips to keep plants bushy. Pruning and harvesting can be synchronous. Prune older plants frequently but lightly. Cut into wood with needles; rosemary won't regrow from bare wood.

HARVEST: Harvest fresh, clean needles. To dry rosemary, bundle sprigs and hang them in a warm, airy place. Once dried, strip off the needles and store in an airtight container. They hold their quality for a year.

NOTES: Rosemary cultivars offer habits that are rounded, erect, or sprawling. Flavor varies with variety; be aware that some varieties are grown more for their form than their flavor. Smaller rosemarys can be grown in pots. Use fast-draining terra cotta and water only as needed when the soil becomes dry. Place the plants in a sunny, south-facing window.

BEST USED IN: liqueurs, beer

Thyme

Thyme is a lovely, evergreen herb that has its place in an evergreen garden, and a culinary one as well. This plant is an easy, well-behaved, shrub-like perennial that you can grow in the herb or vegetable garden, or in a container indoors or out. Another species worth growing is lemon thyme (*Thymus × citriodorus*), an exceedingly handsome golden leaf herb, which will add lemon flavor to a liqueur's flavor.

Use the leaves fresh or dry in small bundles or on drying rack.

If you do any cooking with herbs and spices, you most likely have a bottle of this herb in your spice cabinet. Although typically you don't equate thyme with sweetness, the following simple recipe does just that. Pour one-fifth of vodka (80 to 100 proof) over 2 handfuls of fresh, flowering thyme sprigs. Seal jar and infuse in a cool, dark place for two to eight weeks. Do a weekly taste test until you find the taste of thyme to your liking. Strain the liquid and stir in 1 cup of simple syrup, bottle and seal, and let set for two more months before serving. The more leaves the flowering stems have on it the more pungent the taste.

Thyme

BOTANICAL NAME: *Thymus vulgaris*
PLANT TYPE: woody, evergreen perennial
USDA ZONES: 5–9
HEIGHT: 6–12 inches
SOIL: light, well-drained
LIGHT: full sun. Light shade in hot areas.
WATER: low. Tolerates drought.
GROWTH HABIT: numerous upright, woody stems
PROPAGATE BY: seed, cuttings, layering
SPACING: 12–18 inches
MONTHS TO HARVEST: once plants are large enough to take heavy pruning

PRUNING: Shear mature plants to keep compact.
HARVEST: Harvest the leaves throughout the summer when plants are large enough.
NOTES: Thyme makes a good container plant and may be grown indoors in a sunny window. The plants are usually replaced every few years when they become woody and sparse. Leaves can be used fresh or dried.
BEST USED IN: liqueurs

Acknowledgments

WENDY'S

A toast to my friends and loved ones—Thank you for being exactly who you are. You and me, a well-crafted beverage and time together—life is good!

Thanks in particular to: Sherrye Wyatt of the Northwest Cider Association; Greg Peck, PhD, Assistant Professor of Horticulture at Virginia Tech; and Carol Miles, Professor of Vegetable Horticulture at Washington State University. Special thanks to Shaun Townsend, Assistant Professor of Hop Breeding and Genetics at Oregon State University for generously sharing his expertise, and to Alex Tweten for digging into plants.

DEBBIE'S

To my girls—Kela Calkins and Stephanie Smith.

Thank you to the generous gardeners who opened their gardens for photographs, and shared their knowledge especially Paula Shelkin, Brad and Lori Kittredge, Kris and Cammy Johnson, Jean-Marie DeKoster, Skyler Walker and Alan Fritz, and Terri Mitchell.

Photo Credits

Photos on pages 6, 8, 10, 13, 17, 18, 20, 23, 27, 28, 30, 32, 44, 45, 47, 52, 56, 60, 62, 72, 75, 76, 81, 85, 87, 98, 115, 116, 117, 120, 122, 123, 124, 125, 136, 151, 153, 157, 160, 161, 162, 164, 166, 176, 178, 180, 182, 186, 188, 189, 195, 197, 198, and 201 appear courtesy of Shutterstock.com.

All other photos are from the authors' collections.

Works Cited

Bartholomew, Mel. *All New Square Foot Gardening: Grow More in Less Space!* Nashville, TN: Cool Springs, 2006.

Bartley, Jennifer R. *Designing the New Kitchen Garden: An American Potager Handbook.* Portland, OR: Timber, 2006.

Bennett, Leslie, and Stefani Bittner. *The Beautiful Edible Garden: Design a Stylish Outdoor Space Using Vegetables, Fruits, and Herbs.* Berkeley, CA: Ten Speed, 2013.

Bridle, Bob, ed. *Home Brew Beer.* 1st American Edition ed. New York,NY: DK, 2013.

Buhner, Stephen Harrod. *Sacred and Herbal Healing Beers: The Secrets of Ancient Fermentation.* Boulder, CO: Siris, 1998.

Christensen, Emma. *True Brews: How to Craft Fermented Beer, Wine, Cider, Sake, Soda, Kefir, and Kombucha at Home.* Berkeley, CA: Ten Speed, 2013.

Coupe, R., Roberta Parish, Dennis Lloyd, and Joe Antos. *Plants of Southern Interior British Columbia and the Inland Northwest.* Vancouver, BC: Lone Pine, 1999.

Creasy, Rosalind. *Edible Landscaping.* San Francisco, CA: Sierra Club, 2010.

Crowther, Margaret. *Making Wine with Fruits, Roots & Flowers: Recipes for Distinctive & Delicious Wild Wines.* Cincinnati, OH: Betterway Home/F W Media, 2012.

Doxat, John. *The Indispensable Drinks Book.* New York: Van Nostrand Reinhold, 1981.

Druse, Kenneth. *Making More Plants: The Science, Art, and Joy of Propagation.* New York, NY: Clarkson Potter, 2000.

Eierman, Colby. *Fruit Trees in Small Spaces: Abundant Harvests from Your Own Backyard.* Portland, OR: Timber, 2012.

Fisher, Joe, and Dennis Fisher. *The Homebrewer's Garden: How to Easily Grow, Prepare, and Use Your Own Hops, Brewing Herbs, Malts.* Pownal, VT: Storey, 1998.

Gardner, Jo Ann, and Holly S. Dougherty. *Herbs in Bloom: A Guide to Growing Herbs as Ornamental Plants.* Portland, OR: Timber, 1998.

Hamilton, Andy. *Booze for Free: The Definitive Guide to Making Beer, Wines, Cocktail Bases, Ciders, and Other Drinks at Home.* New York, NY: Plume, 2013.

Hill, Lewis, and Leonard P. Perry. *The Fruit Gardener's Bible: A Complete Guide to Growing Fruits and Berries in the Home Garden*. Pownal, VT: Storey, 2012.

Hughes, Greg. *Home Brew Beer*. New York, NY: DK, 2013.

Kershaw, Linda, A. MacKinnon, and Jim Pojar. *Plants of the Rocky Mountains*. Edmonton, AB: Lone Pine Pub., 1998.

Kershaw, Linda. *Edible & Medicinal Plants of the Rockies*. Edmonton, AB: Lone Pine Pub., 2000. Print.

Kowalchik, Claire, William H. Hylton, and Anna Carr. *Rodale's Illustrated Encyclopedia of Herbs*. Emmaus, PA: Rodale, 1987.

Lima, Patrick, and Turid Forsyth. *Herbs: The Complete Gardener's Guide*. Willowdale, ON: Firefly, 2001.

McCormick, Ann. "The Art of the Artemisia." *Herb Quarterly* Summer 2014: 29–33.

McGee, Rose Marie Nichols, and Maggie Stuckey. *The Bountiful Container: A Container Garden of Vegetables, Herbs, Fruits, and Edible Flowers*. New York, NY: Workman, 2002.

Meyer, Scott. *Hooch: Simplified Brewing, Winemaking, and Infusing at Home*. Philadelphia, PA: Running, 2013.

Miodownik, Mark. *Stuff Matters: Exploring the Marvelous Materials That Shape Our Man-made World*. New York, NY: Houghton Mifflin Harcourt, 2014.

Moerman, Daniel E. *Native American Ethnobotany*. Portland, OR: Timber, 1998.

Morgan, Joan, Elisabeth Dowle, and Alison Richards. *The New Book of Apples: The Definitive Guide to Apples, Including over 2000 Varieties*. n.p.: Ebury, 2003.

Murphy, Kevin. "Springtime Cocktails from the Garden." *Herb Quarterly* Spring 2014: 36–39.

National Center for Biotechnology Information. "Rhododendron Tomentosum (Ledum Palustre). A Review of Traditional Use Based on Current Research." U.S. National Library of Medicine, n.d., accessed May 26, 2014, www.ncbi.nlm.nih.gov/pubmed/23352748.

Normandeau, Sheryl. "Herb-Infused Shrub Syrups." *Herb Quarterly* Summer 2014: 39–42.

North Carolina State University Cooperative Extension "Juniperus Communis," n.d., accessed July 27, 2014, plants.ces.ncsu.edu/plants/all/juniperus-communis/.

Ortiz, Elisabeth Lambert. *The Encyclopedia of Herbs, Spices & Flavorings.* New York, NY: Dorling Kindersley, 1992.

Papazian, Charlie. *The Complete Joy of Homebrewing.* 3rd ed. New York, NY: William Morrow, 2003.

Peragine, John N. *The Complete Guide to Growing Your Own Hops, Malts, and Brewing Herbs: Everything You Need to Know Explained Simply.* Ocala, FL: Atlantic Publishing Group, 2011.

Pitzer, Sara. *Homegrown Whole Grains: Grow, Harvest & Cook Wheat, Barley, Oats, Rice, Corn & More.* North Adams, MA: Storey Pub., 2009.

Plants For A Future. "Juniperus Communis Juniper, Common Juniper." *PFAF Plant Database*, n.d., accessed July 27, 2014, www.pfaf.org/user/Plant. aspx?LatinName=Juniperus+communis.

Plants For A Future. "Ledum Palustre Wild Rosemary, Marsh Labrador Tea." *PFAF Plant Database*, n.d., accessed Novenber 6, 2014. www.pfaf.org/user/Plant. aspx?LatinName=Ledum+palustre.

Pollan, Michael. *The Botany of Desire: A Plant's Eye View of the World.* New York, NY: Random House, 2001.

Pooley, Michael, and John Lomax. *Real Cidermaking on a Small Scale: An Introduction to Producing Cider at Home.* East Petersburg, PA: Fox Chapel Pub., 2011.

Proulx, Annie, Lew Nichols. *Cider: Making, Using & Enjoying Sweet & Hard Cider.* Third ed. Pownal, VT: Storey Communications, 2003.

Rainy Side Gardeners. "Tips for the Garden." Rainy Side Gardeners, n.d., accessedSeptember 10, 2014, www.rainyside.com/features/ GardenTips.html.

Schloss, Andrew. *Homemade Liqueurs and Infused Spirits.* N.p.: n.p., n.d.

Solomon, Steve. *Growing Vegetables West of the Cascades: The Complete Guide to Organic Gardening.* Seattle, WA: Sasquatch, 2007. Print.

Stewart, Amy. *The Drunken Botanist: The Plants That Create the World's Great Drinks.* Chapel Hill, NC: Algonquin of Chapel Hill, 2013. Print.

Schloss, Andrew. *Homemade Liqueurs and Infused Spirits.* North Adams: Storey, 2013. Print.

Seiler, John. "Myrica Gale Fact Sheet." *Myrica Gale Fact Sheet*. Virginia Tech Department of Forest Resources, n.d. Web. 06 Nov. 2014. dendro.cnre. vt.edu/dendrology/syllabus/factsheet. cfm?ID=436>.Solomon, Steve. *Growing Vegetables West of the Cascades: The Complete Guide to Organic Gardening*. Seattle, WA: Sasquatch, 2007. Print.

Stewart, Amy. *The Drunken Botanist: The Plants That Create the World's Great Drinks*. Chapel Hill, NC: Algonquin of Chapel Hill, 2013. Print.

Stika, John. "Malting Your Own: Techniques." *Malting Your Own: Techniques*. Brew Your Own, n.d. Web. 10 June 2014. byo.com/ stories/item/1108-malting-your-own-techniques>.

Tirmenstein, D. "Species: Juniperus Communis." U.S. Forest Service, n.d., accessed July 29, 2014, www.fs.fed.us/ database/feis/plants/shrub/juncom/ all.html.

Ultimate Guide to Fences, Arbors, and Trellises: Plan, Design, Build. Upper Saddle River, NJ: Creative Homeowner, 2008. Print.

USDA, ARS, National Genetic Resources Program. "Triticum Aestivum Subsp. Spelta Information from NPGS/ GRIN." *Triticum Aestivum Subsp. Spelta Information from NPGS/GRIN*. Germplasm Resources Information Network - (GRIN) [Online Database]. National Germplasm Resources Laboratory, Beltsville, Maryland., n.d. Web. 10 June 2014. www.ars-grin.gov/ cgi-bin/npgs/html/taxon.pl?406903>.

Virginia Tech—Department of Forest Resources and Environmental Conservation. "Juniperus Communis Fact Sheet." n.d., accessed July 28, 2014, dendro.cnre.vt.edu/dendrology/ syllabus/factsheet.cfm?ID=212.

Whitten, Greg. *Herbal Harvest: Commercial Organic Production of Quality Dried Herbs*. 3rd ed. Melbourne, Australia: Bloomings, 2004.

Wilder, Louise Beebe. *The Fragrant Path; a Book about Sweet Scented Flowers and Leaves*. Vancouver, BC: Hartley & Marks, 1996.

Wendy Tweten is an award-winning writer and speaker who lives and gardens (and occasionally imbibes) on the Kitsap Peninsula in the Puget Sound. Along with her alter-ego, Miss Snippy, she contributes to a number of national and regional publications and websites. She is a regular columnist for the Kitsap News Group family of publications, and her work as been featured in *Organic Gardening*, *Northwest Garden News*, *Master Gardener* magazine, and *Home by Design* among others. She has won four Garden Writers' Association silver awards and two gold awards for her writing. Wendy's philosophy is simple: Design, color theory, and impressing the neighbors aside, the true function of a garden is to entertain the gardener.

Debbie Teashon is a freelance garden writer, author, and award-winning photographer based on the Kitsap Peninsula in Washington. She has gardened most of her adult life and written about it for over two decades and her photography career spans four decades. Debbie's articles and photographs have appeared in magazines such as *Fine Gardening*, *Master Gardeners*, *West Sound Home and Garden*, *Master Gardeners*, and *The Oregonian*, among others. As a plantswoman, she spends her time gardening, taking classes or researching plants for articles and the online plant database she maintains on Rainy Side Gardeners (www.rainyside.com), a website to help gardeners in the Pacific Northwest.